Springer Series in
MATERIALS SCIENCE 128

Springer Series in
MATERIALS SCIENCE

Editors: R. Hull   R. M. Osgood, Jr.   J. Parisi   H. Warlimont

The Springer Series in Materials Science covers the complete spectrum of materials physics,
including fundamental principles, physical properties, materials theory and design. Recognizing
the increasing importance of materials science in future device technologies, the book titles in this
series reflect the state-of-the-art in understanding and controlling the structure and properties
of all important classes of materials.

Please view available titles in *Springer Series in Materials Science*
on series homepage http://www.springer.com/series/856

Victor I. Mikla
Victor V. Mikla

Metastable States in Amorphous Chalcogenide Semiconductors

With 68 Figures

 Springer

Prof. Dr. Victor I. Mikla
Uzhgorod State University
Voloshina str., 54
Uzhgorod 88000
Ukraine
E-mail: mikla@univ.uzhgorod.ua

Dr. Victor V. Mikla
Chelyskincev str., 6/62
Uzhgorod 88000
Ukraine
E-mail: mikla_v@ukr.net

Series Editors:

Professor Robert Hull
University of Virginia
Dept. of Materials Science and Engineering
Thornton Hall
Charlottesville, VA 22903-2442, USA

Professor R. M. Osgood, Jr.
Microelectronics Science Laboratory
Department of Electrical Engineering
Columbia University
Seeley W. Mudd Building
New York, NY 10027, USA

Professor Jürgen Parisi
Universität Oldenburg, Fachbereich Physik
Abt. Energie- und Halbleiterforschung
Carl-von-Ossietzky-Straße 9–11
26129 Oldenburg, Germany

Professor Hans Warlimont
DSL Dresden Material-Innovation GmbH
Pirnaer Landstr. 176
01257 Dresden, Germany

Springer Series in Materials Science   ISSN 0933-033X
ISBN 978-3-642-26159-6   e-ISBN 978-3-642-02745-1
DOI 10.1007/978-3-642-02745-1
Springer Heidelberg Dordrecht London New York

Cover design: SPi Publisher Services

Printed on acid-free paper

Springer is part of Springer Science+Business Media (www.springer.com)

Preface

This monograph deals with metastable states in amorphous semiconductors – materials which lack long-range periodicity in the atoms' positions, which are in thermodynamic nonequilibrium and which, in addition, have several metastable states. These states give rise to various properties and effects – namely a wide range of photoinduced changes and high photosensitivity and X-ray sensitivity – that are unique among solid-state semiconductors. Historically, amorphous selenium and selenium-based materials have played an important role in physics and technology, and they continue to do so. In these materials there exist inherent intermediate (metastable) states, structural and electronic in origin, which lead to interesting properties and effects different from those of their crystalline counterparts.

In this volume, the metastable states and related effects are investigated in depth against the background of a detailed consideration of local atomic and electronic structure, and taking into account a wide range of light-induced effects.

Although the first publications on amorphous semiconductors date back to the early 1970s, studies of metastable states in these materials had not been analyzed systematically up to now, which led to erroneous ideas, even among specialists. In the present book, experimental investigations of metastable states are reported in detail for elemental selenium and selenium-based materials.

This monograph thus represents a complete course on metastable states in selenium and selenium-based materials, which may be of interest to different groups of readers. On the other hand, it is also intended as a reference book for scientists and engineers who are actively involved in amorphous semiconductor material characterization and defect engineering in research institutes and industrial laboratories. Chapter summaries and tables compiling the theoretical, methodological and experimental results allow quick access to the major results and simple location of detailed information in the text. The data evaluation technique presented here provides insight into the analytical potential of the various techniques, for example Raman scattering, thermally stimulated depolarization currents, and time-of-flight xerographic spectroscopy.

Concerning manuscript preparation, I am grateful to Dr. Angela Lahee for her helpful comments. We would like to thank the language editor, Deborah Hollis, for her very expert work. Finally, I would like to express my deepest thanks to my wife, Ottilia, for her unfailing support and patience, and for understanding that compiling a monograph requires much time.

Uzhgorod   Victor I. Mikla
June 2009

Contents

Acronyms and Abbreviations/Symbols

CTL   Charge transport layer
dc   Direct current
EXAFS   Extended X-ray absorption fine structure
FSDP   First sharp diffraction peak
FTIR   Fourier transform infrared
IR   Infrared
IVAP   Intimate valence alternation pair
Nd:YAG   Neodymium-doped yttrium aluminium garnet
PD   Photodarkening
PGL   Photogeneration layer
PIDC   Photoinduced discharge curve
PMMA   Poly(methylmethacrylate)
TOF   Time-of-flight
TSC   Thermally stimulated conductivity
TSDC   Thermally stimulated depolarisation current
VAP   Valence alternation pair
XRD   X-ray diffraction
2D   Two-dimensional

C Capacitance
d Sample thickness
E_g Energy gap
E_g^T Tauc gap
E_t Thermal activation energy
E_{th} Threshold energy density
E_σ Conductivity activation energy
G Conductance
h Planck constant
I Intensity
I_{ph} Photocurrent
J_D current density

k	Boltzmann constant
N_v	Density of states in the valence band
n	Refractive index
P	Power density
p	Concentration of free holes in the valence band
Q_0	Initial charge
R	Reflectance
R_c	Structural correlation length
S	Capture cross section
T	Temperature; transmission
T_g	Glass transition temperature
T_{rel}	Relative transmissivity
t	Time
t_T	Transit time
U	Voltage
U_r	Residual voltage
V	Velocity of sound; voltage
v_T	Heating rate
α	Absorption coefficient
ε	Dielectric constant
$\zeta(\tau)$	Relaxation function
η	Diffraction efficiency
Θ	Angle
Λ	Grating pitch
λ	Wavelength
μ	Mobility
σ	dc conductivity
τ	Dark-resting time
τ_{MR}	Monomolecular recombination time
χ	Permittivity
$\psi(\tau)$	Relaxation function
ω	Frequency [cm^{-1}]
ω_{max}	Frequency of the peak maximum

Chapter 1
Introduction

Investigation of noncrystalline semiconductors is one of the most interesting and attractive disciplines of condensed matter physics. In fact, these investigations are strongly stimulated by both scientific and technological factors. Such materials are free of limitations inherent to their crystalline counterparts with long-range order of the positions of the atoms. They have a wide range of applications as given below.

The intellectual attractiveness of noncrystalline semiconductors is explained by new approaches. The latter are not connected with the terminology of the Brillouin zone, selection rules and steric arguments, which are not applicable to disordered (noncrystalline) systems. Another reason that makes new approaches necessary is connected with the absence of long-range order. In fact, most conventional techniques typically used for classical crystalline semiconductors are complicated to realize at conditions in which the carrier range is of the order of interatomic distances and/or when the interpretation of experimental data is ambiguous. In addition, conventional spectroscopic techniques give structureless spectra without clear features that would help in understanding fundamental properties.

Amorphous (vitreous) chalcogenides are materials containing one or more chalcogenide elements (group VI in the periodic table, e.g. sulfur, selenium or tellurium) as a substantial constituent. They are covalently bonded materials and may be classified as a molecular solid. It is important to emphasise that they behave like semiconductors.

The unique properties of this class of materials are explained by structural disorder, thermodynamic nonequilibrium and metastability. The designation *metastable state* is reserved for states whose lifetimes are relatively long. Metastability is a general scientific concept that describes states of delicate equilibrium. A system is in a metastable state when it is in equilibrium but is susceptible to fall into a lower-energy state with only slight interaction. This may be bandgap illumination, and/or annealing. A metastable state is thus considered a somewhat stable intermediate stage of a system, the energy of which may be lost in discrete amounts. For example, many chalcogenides exhibit a light-driven metastability called photodarkening. Photodarkening is a reduction in bandgap on exposure to light. The physical origin of the process, which takes place during various structural transformations induced by bandgap light and thermal annealing, remains unclear. It is necessary to note that quasi-stationary states – the initial and final, after the corresponding

V.I. Mikla and V.V. Mikla, *Metastable States in Amorphous Chalcogenide Semiconductors*, Springer Series in Materials Science 128, DOI 10.1007/978-3-642-02745-1_1, © Springer-Verlag Berlin Heidelberg 2010

transformation – are the preferentially studied ones. At the same time, metastable states, inherent to transition processes in the system, are poorly studied. Metastable states may be preferentially *structural* (on the atomic and molecular scale) or *electronic*. Here it is necessary to note that transitions between metastable states is difficult to avoid (in other words entirely "pure" experiments are unrealistic). At the same time, the presence of metastable states causes unique effects and gives rise to very interesting physical properties. Questions about the structure of amorphous chalcogenides on the scale of medium-range order and possible modification of this structure by variation of composition, technological factors and subsequent treatment (bandgap irradiation or annealing) remain unanswered. It is also necessary to obtain a deeper insight into the electronic structure of states in the mobility gap, as well as the possibility of controlling these states.

Even amorphous materials have some short-range order at the atomic length scale owing to the nature of chemical bonding. Furthermore, in very small crystals a large fraction of the atoms are located at or near the surface of the crystal; relaxation of the surface and interfacial effects distort the atomic positions, decreasing the structural order. Even the most advanced structural characterization techniques, such as X-ray diffraction and transmission electron microscopy, have difficulty in distinguishing between amorphous and crystalline structures on these length scales. The only helpful and very informative technique in this context is, to the authors' knowledge, light scattering (Raman) spectroscopy. This technique is powerful and probes structural metastable states. Electronic metastable states are "visualized" by means of thermally stimulated depolarization currents, time-of-flight and electrophotographic (xerographic) measurements.

Raman scattering is the inelastic scattering of photons. When light is scattered from an atom or molecule, most photons are elastically scattered, such that the scattered photons have the same energy (frequency) and wavelength as the incident photons. However, a small fraction of the scattered light is scattered by an excitation, with the scattered photons having a frequency different from the frequency of the incident photons. Raman scattering is a significant tool for analyzing the molecular structure of amorphous semiconductors.

Thermally stimulated depolarization allows the shallow metastable electronic states in the mobility gap of amorphous materials to be accessed. In this method the semiconductor sample is previously polarized at low temperature, then, in the process of heating, nonequilibrium carriers are liberated from metastable states.

In general, time-of-flight (TOF) describes a method in which the time that it takes for a particle (e.g. charge carrier) to reach a detector is measured when the particle travels a known distance (in solids, the sample thickness).

Xerographic spectroscopy is based on several stages of the well-known xerographic process. In this technique a metallic plate coated with, for example, amorphous selenium is used. Amorphous selenium will hold an electrostatic charge in darkness and will conduct away such a charge under light. There are two important steps in xerographic measurements. The first step is charging. An electrostatic charge is uniformly distributed over the surface of the photosensor (selenium layer) by a corona discharge. The next step is illuminating the charged photosensor. In this step the charge will be dissipated. From the corresponding discharge characteristics the states in the mobility gap may be extracted.

Detailed descriptions of the above techniques are given in the corresponding chapters.

Chalcogenide glasses have great potential for applications in photonics. Four kinds of applications are commercially available or used practically [1]. These rely upon the unique features of chalcogenide glasses: quasistability (metastability), photoconductive properties, infrared transparency and ionic conduction.

The first is the phase-change phenomenon used in erasable high-density optical memories [2]. These use semiconductor lasers and chalcogenide films such as GeSbTe with a thickness of approximately 15 nm. The reflectivity change between amorphous and crystalline phases is monitored with a weak light beam. The present memory capacity is about 5 GB/disk and is still increasing.

The second category is photoconductive applications such as photoreceptors in copying machines and X-ray imaging plates [3]. In the photoconductive target in vidicons (type of video camera), avalanche multiplication in amorphous selenium films is employed, which substantially enhances the light sensitivity so that star images can be taken.

The third application is purely optical [4]. That is, since the chalcogenide glass is transparent in the IR region, it can be utilized for IR optical components such as lenses and windows. It can also be utilized for IR-transmitting optical fibers. Fibers have also been employed as a matrix to incorporate rare-earth ions. Such rare-earth ion doped chalcogenide glasses are promising for the preparation functional fibers such as optical amplifiers.

Lastly, chalcogenide glasses containing group I elements such as silver are used as high-sensitivity ionic sensors [4]. Some lithium-containing glasses have also been utilized as solid-state electrolytes in all-solid batteries.

It is important to note that more effective application of noncrystalline materials for optical data storage, in holography, xerography and radiography depends strongly on our knowledge of molecular and electronic structure. All of the arguments mentioned are reasons for the current importance of experimental studies of metastable states and relaxation effects in amorphous chalcogenide semiconductors.

Selenium is one of the most important and intriguing representatives of a wide class of noncrystalline chalcogenide semiconductors. It is a very promising material as a nearly "ideal" model object for fundamental studies. The structure of amorphous selenium (a-Se) is formed preferentially by chain fragments. Intra-atomic bonding is stronger than interatomic bonding. Therefore, semiconducting amorphous selenium appears to be a molecular material and may be considered in some sense a polymer. Addition of arsenic or antimony to amorphous selenium makes it possible to observe the influence of these additives on the specific photophysical properties. At the same time, one can observe the transition from one-dimensional (chain-like) structure to two-dimensional (layer-like) structure. In this book elemental selenium and $As\,(Sb)_x\,Se_{1-x}$ alloys are chosen as model objects because the experimental results obtained are common to a wide range of noncrystalline chalcogenide semiconductors. The physical origin of all of the unique properties observed, the authors believe, lies in the fact that amorphous semiconductors are structurally disordered, thermodynamically nonequilibrium and kinetically unstable materials. It should be pointed out also that in this book the conventional approach dealing with structure, fundamental electronic properties and their explanation in terms of defects is not followed. The authors have tried to emphasise only intriguing structural and electronic metastable states in the mobility gap of amorphous chalcogenide semiconductors. The authors leave it to the reader to decide to what extent such an approach is successful.

References

1. K. Tanaka, Chalcogenide Glasses, in *Encyclopedia of Materials: Science and Technology* (Elsevier, Oxford, 2001), pp. 1123–1131
2. A. Madan, M.P. Shaw, *The Physics and Applications of Amorphous Semiconductors* (Academic, Boston, MA, 1988)
3. S.O. Kasap, J.A. Rowlands, J. Mater. Sci. Mater. Electron. **11**, 179 (2000)
4. A. Feltz, *Amorphous Inorganic Materials and Glasses* (VCH, Weinheim, Germany, 1993)

Chapter 2
Effect of Thermal Evaporation Conditions on Structure and Structural Changes in Amorphous Arsenic Sulfides

In this first chapter we consider, using a powerful technique such as Raman scattering, structural changes induced by light and heat treatment in vacuum-deposited amorphous chalcogenide films prepared by thermal evaporation at different rates. The changes in the Raman spectra of thermally evaporated amorphous As_2S_3 films after light treatment are interpreted in terms of rearrangement of bonding configurations of molecular species that exist just after evaporation. As the evaporation temperature or deposition rate is increased, irradiation with bandgap light, As-enriching and/or further annealing of the films induces the formation of microcrystallites. At the same time, the reversible photoinduced changes in annealed films involve negligibly small changes in bonding statistics. Discernable change occurs in the medium-range order.

2.1 Introduction to Amorphous Solids

An amorphous solid is a solid in which there is no long-range order of the positions of the atoms. Solids in which there is long-range atomic order are called crystalline. Most classes of solid materials can be prepared in an amorphous form. For instance, chalcogenide glasses are amorphous materials.

In principle, given a sufficiently high cooling rate, any liquid (melt) can be made into an amorphous solid. If the cooling rate is faster than the rate at which molecules can organise into a more thermodynamically favourable crystalline state, an amorphous solid will be formed. In contrast, if molecules have sufficient time to organise into a structure with two- or three-dimensional (2D or 3D) order, a crystalline solid will be formed.

Some materials, such as metals, are difficult to prepare in an amorphous state. Unless a material has a high melting temperature or a low crystallisation energy, cooling must be done extremely rapidly. As cooling is performed, the material changes from a supercooled liquid, with properties one would expect from a liquid-state material, to a solid. The temperature at which this transition occurs is called the glass transition temperature T_g.

V.I. Mikla and V.V. Mikla, *Metastable States in Amorphous Chalcogenide Semiconductors*, Springer Series in Materials Science 128, DOI 10.1007/978-3-642-02745-1_2, © Springer-Verlag Berlin Heidelberg 2010

An amorphous solid, such as a chalcogenide glass, is a rigid material whose structure lacks crystalline periodicity. Thus the pattern of its constituent atoms or molecules does not repeat periodically in three dimensions. In the present terminology amorphous and noncrystalline are synonymous. A solid is distinguished from its other amorphous counterparts, that is, liquids and gases, by its viscosity.

The structural, optical and photophysical properties of chalcogenide glasses have been the subject of permanent systematic interest for more than 30 years [1–4]. The interest has been stimulated both by unresolved fundamental scientific problems (their structure and electronic properties, to mention just two) of noncrystalline materials and the need to assess their potential as optical memory elements, for image formation and other useful applications.

The structural and physical properties of amorphous chalcogenides depend on the preparation conditions and subsequent sample treatments, although in a less sensitive manner than their crystalline counterparts. In many cases the initial structures of amorphous films differ from those of well-annealed films or their bulk glasses, and they undergo significant irreversible structural changes in response to band-gap illumination or annealing (so-called photo- and thermo-induced changes). Although the induced effects mentioned above have been reviewed many times, their origin remains disputable.

In the following we attempt to present our experience in examining the influence of preparation conditions (e.g. thermal evaporation rates) and of various treatments (irradiation or annealing) on the structure of fresh (as-evaporated) $As_x S_{1-x}$ amorphous films. The latter are particularly suitable objects for such a study because the irreversible changes in the optical properties of these materials, especially in $As_2 S_3$, are much more pronounced than those in other chalcogenide glasses, e.g. $As_x Se_{1-x}$.

2.2 Experimental Investigation of the Influence of Thermal Evaporation Conditions

Usually the samples used in these studies were amorphous films, typically 5–20 μm thick. The films were prepared in conventional manner, by thermal vacuum evaporation of glassy $As_x S_{1-x}$ alloys (high-purity elements were alloyed by melting in evacuated and sealed quartz ampoules) from a molybdenum boat onto glass and quartz substrates. The evaporation was performed under two different conditions: evaporation onto a substrate with deposition rate ca. $20 \, \text{Å} \, \text{s}^{-1}$ (conventional mode) and evaporation onto a substrate with deposition rate ca. $1,000 \, \text{Å} \, \text{s}^{-1}$ (flash evaporation). The deposition rate was varied by changing the source (evaporation boat) temperature while the substrate temperature was held constant at $T = 300$ K. Chemical compositions of the films, e.g. $As_2 S_3$, were found to be $As_{39.7} S_{40.3}$ on the basis of electron microprobe analysis. The lack of crystallinity in the film samples was systematically verified by X-ray diffraction measurements.

Raman spectra were measured by conventional and Fourier transform IR (FTIR) Raman spectrometry (Bruker, model IFS 55). Laser irradiation of wavelengths

647.1 and 1,064 nm from Kr-ion and neodymium-doped yttrium aluminium garnet (Nd:YAG) lasers were used for the excitation of the Raman spectra. Such a low energy, especially in the latter case, is well below the values of the optical bandgaps for the glassy alloys under study. Thus, no detectable photostructural transformation took place under data collection. It is quite natural that the amplitudes of some Raman lines from As_xS_{1-x} films were often low (compared to that in bulk glass); the spectra were usually "noisy" and not smooth. The backscattering method was used with a resolution of $1\,cm^{-1}$. Raman spectra of the amorphous films were recorded at sufficiently low incident laser-beam power density (3–5 mW) to avoid possible photostructural changes in the case of the 647 nm excitation line. A Kr^+-ion laser was used as an exposure source and the films were exposed at an intensity of $30\,mW\,cm^{-2}$. Although structural transformations in As_xS_{1-x} films have been the subject of several articles [5–17], some important aspects of them are not still fully understood.

First of all we consider the results obtained for conventionally prepared amorphous films. Figure 2.1 shows the Raman spectra for the as-deposited (1), exposed (2), annealed film (3), and bulk-glass (4) forms of As_2S_3.

Qualitatively similar results for irreversible thermostructural transformation have been firstly reported by Solin and Papatheodorou [5], Nemanich et al. [6], Mikla et al. [7], and Frumar [8, 9]. The spectrum of the as-deposited film consists of relatively sharp peaks on the background, typical of bulk As_2S_3 glass. These sharp features, characteristic of the as-deposited film, broaden, decrease in intensity or even disappear irreversibly after annealing or on illumination. Such a behaviour

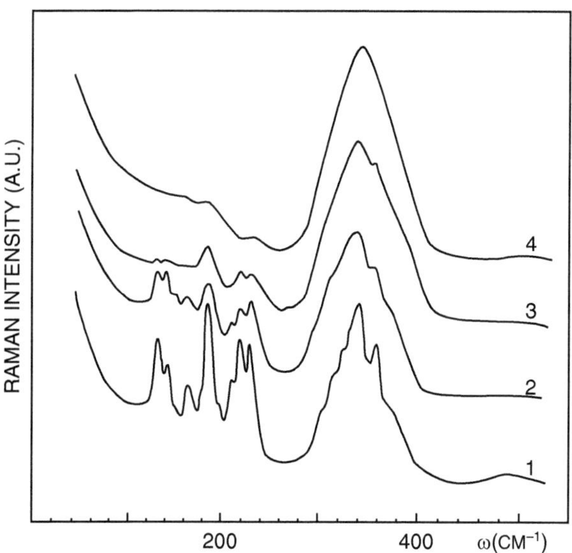

Fig. 2.1 Raman spectra of as-deposited (1), photodarkened (2), and annealed (3) As_2S_3 amorphous films. For comparison, the spectrum (4) of bulk glass is also shown

indicates that significant (in the sense of amorphous semiconductors) structural modification has taken place. Note that in bulk glass and amorphous film form with As content exceeding 40 at% all the relatively sharp features mentioned above are also observed. The data analysis [7–11] indicates that the sharp spectral features (except for the $490\,cm^{-1}$ band) originate from "wrong" (like-atom or homopolar) As–As bonds in As_4S_4 molecular units. The presence of As_4S_4 molecules in the as-deposited As_2S_3 films is strongly supported by IR [12], EXAFS [13] and recent Raman scattering [7–9] measurements. Taking this into account, as-deposited films of As_2S_3 possess AsS_3 pyramidal units, which form a glassy matrix of As_2S_3 and, in addition to them, a partially polymerised mixture of As_4S_4 (realgar-type) and S_n molecular species.

After annealing, the spectrum of the as-deposited film becomes very similar to that of bulk glass. However, some broad features – the maxima at 187 and $230\,cm^{-1}$ – remain. This indicates that some "wrong" bonds of the type As–As and S–S exist even in well-annealed As_2S_3 films. During this treatment the bond breaking and switching due to an increased mobility of atoms is accompanied (to a considerable extent) by structure polymerisation. This is manifested in the S–As–S stretching mode $(120–170\,cm^{-1})$ frequency spreading and vibration band broadening typical of the bulk glass. Similar behaviour in the valence mode region of the As–S and As–As vibration bands takes place if we illuminate the as-deposited film. Consequently, we can say that the structure of illuminated films remains close to that of as-deposited films.

In most studies on arsenic chalcogenide films, including those examined in the work discussed here, the investigators deposited amorphous layers at fairly low deposition rates ranging maximally from 10 to $100\,\text{Å}\,s^{-1}$. The reaction of these samples to bandgap illumination is seen as an absorption edge shift to lower energies (red shift or photodarkening, PD). In contrast, setting the evaporator at $T_{evap} = 800°C$ to $900°C$ and deposition rates of $\geq 300\,\text{Å}\,s^{-1}$ lead to a structural modification, which in turn results in a decrease in the optical bandgap. The exposed samples behave like "positive" photosensitive layers (in contrast to materials with negative image formation), that is, they exhibit a blue shift of the absorption edge. In Fig. 2.2, a typical Raman spectrum of a film deposited at a higher rate is shown.

In general, it is clear that on raising the evaporation temperature and deposition rate the spectral bands sharpen and the smooth background disappears. The high-frequency component at $490\,cm^{-1}$, which represents S–S bonds polymerised in the glass network, disappears. This is paralleled by the appearance of additional features at 120, 150, 200, 233 and $273\,cm^{-1}$. Note that the strongest peak $(273\,cm^{-1})$ in the Raman spectrum of the film deposited at high deposition rate is absent in those of As_2S_3 crystals, and α, β-As_4S_4 polymorphs. Recall that this line was observed [14] in the Raman spectrum of the sample of crystalline As_4S_3 obtained by vacuum sublimation of melted As_4S_3.

As the deposition rate increased, the films were found to become sulfur-deficient. This is the case even for sulfur-rich compositions (see Fig. 2.3). One of the reasons for such As-enriching of the samples may be the partial fragmentation of As_2S_3 into As_2S_2 and sulfur during the deposition. Note that the increase in quenching

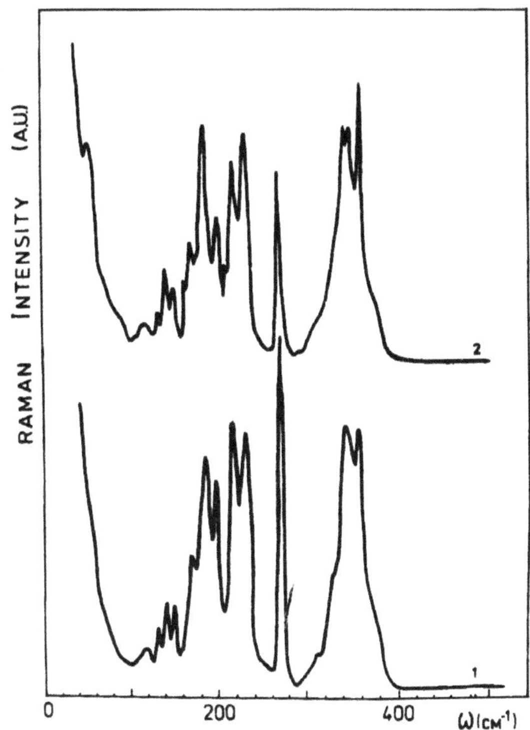

Fig. 2.2 Raman spectra of a fast-deposited As_2S_3 film. Spectrum 1 was taken before annealing and spectrum 2 after annealing

temperature has a similar effect on the structure of melt-quenched As_2S_3, as was demonstrated by Yang et al. [13], Mikla et al. [7] and Zitkovsky and Boolchand [15].

Only a slight change in the Raman spectrum was observed after irradiation of such films. In contrast, annealing resulted in further narrowing of the bands, splitting and intensity redistribution. The low-frequency region of the spectra of annealed films exhibits a distinct shoulder at $60\,cm^{-1}$, which probably corresponds to the lattice (intermolecular) line in the spectra of the crystals α, β-As_4S_4 [11].

It is important to point out here that the transformation of the Raman spectrum with increasing As content (above stoichiometric $As_{0.4}S_{0.6}$ composition) is similar to that observed for As_2S_3 with the increase in deposition rate (compare Figs. 2.2 and 2.4). Results suggest that the high deposition rate induces the formation of microcrystallites. Annealing further enhances the crystallisation processes and the films become polycrystalline. Such a radical structural transformation, observed in the films deposited at high rates, was found to be irreversible – the initial structure (and the corresponding absorption edge position) could not be restored even by annealing near T_g. Moreover, the films deposited at higher rates always appeared smooth, but took on a "dusty" appearance after light and, especially, heat treatment.

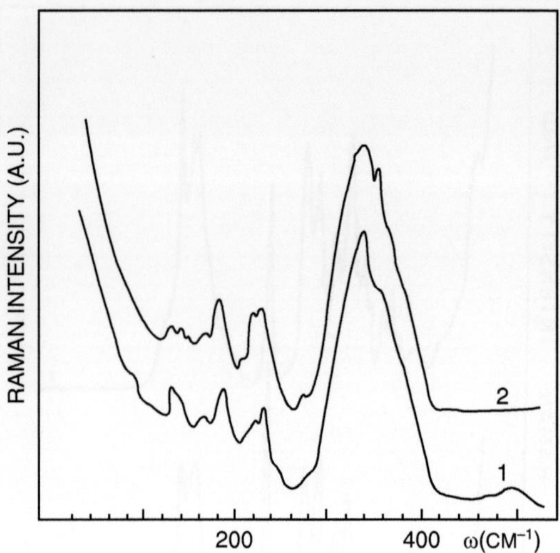

Fig. 2.3 Raman spectra of $As_{0.30}S_{0.70}$ films. *Curve* 1 is the spectrum of the film prepared in conventional manner (slow deposition); *curve* 2 corresponds to the fast-deposited film

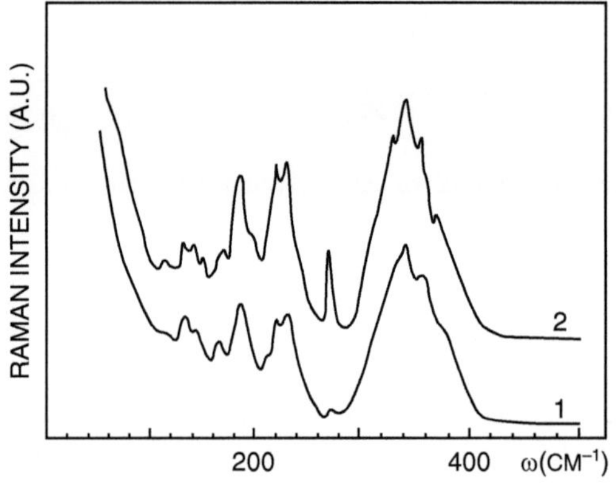

Fig. 2.4 Raman spectra of $As_{0.44}S_{0.56}$ films prepared by slow (1) and fast (2) evaporation

Returning to the films prepared in conventional mode (slow-deposition films), we next examine how PD of well-annealed films and bulk samples influences their structure. In this case, the irreversible changes are much less developed, principally because the magnitude of the reversible PD is small. Structural changes detected by means of direct structural probes (e.g. X-ray diffraction [16]) were subtle. Regrettably, indirect probes of local structure by means of Raman spectroscopy, even in the case of Fourier transform Raman spectroscopy, which is believed to be more sensitive than conventional Raman spectroscopy, also shows only negligibly small changes between annealed and irradiated amorphous films, as shown by the data presented here. At the same time, it should be noted that Raman scattering is sensitive

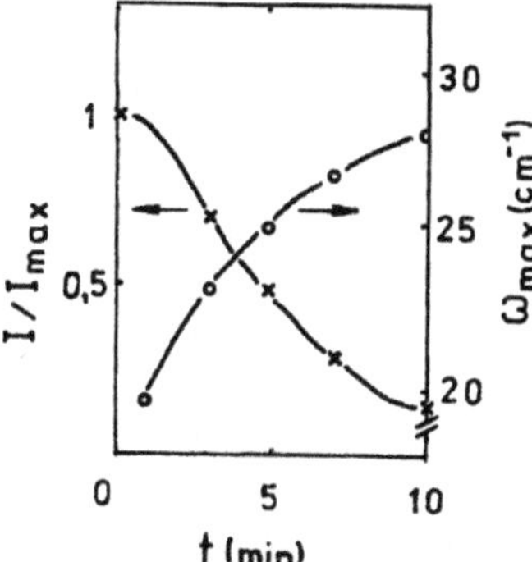

Fig. 2.5 The dependence of the low-frequency peak position and relative intensity on exposure time. I_{max} denotes the peak intensity for a nondarkened sample

to medium-range order and can provide some structural information in terms of correlation length. In [8] Frumar et al. reported an increase in the intensity of the $231\,\text{cm}^{-1}$ Raman band characteristic of As–As bonds in a-As_2S_3 with PD. In contrast, in the present experiments not even a slight change in Raman spectra in the frequency range 100–$500\,\text{cm}^{-1}$ could be detected. Only low-frequency regions of the Raman spectra show discernible changes under illumination.

Figure 2.5 summarises the low-frequency Raman spectra measurements at room temperature. As_2S_3 glassy samples were illuminated for different times from 1 to 240 min. It may be observed that the peak position of the spectrum gradually shifts to higher frequencies as the illumination time is increased. The reversible PD in As_2S_3 annealed thin films and bulk glasses are also accompanied by changes in the intensity of the low-frequency peak. It is clear that the PD has appreciably reduced the intensity of the boson peak.

The irreversible structural changes that occur in the conventionally prepared (slowly deposited) thin films upon optical illumination can be understood in terms of the mechanism described in [5–12, 17]. These changes are associated with a photopolymerisation of As_4S_4 and S_n molecular species. The resultant structure after such a transformation is 2D, layer-like, with nearly restored chemical ordering.

Rapid deposition, as observed in the Raman data, significantly enhances compositional as well as structural disorder in amorphous samples compared with that of the stoichiometric composition. An increase in As concentration leads to a condensation of As_4S_3 and some other As-rich molecular species; a disordered molecular solid is obtained. These As-rich compositions are unstable and presumably phase separation occurs upon thermal annealing and light irradiation. Without going further into the details of these irreversible effects in fast-deposition films, it may be suggested that irreversible photoinduced structural changes in conventionally prepared amorphous films are quite different from a phenomenological point of view and also with respect to their origin.

On the basis of the Raman results presented here, one can conclude that, in contrast to the irreversible structural changes, no changes in local structure appear to accompany the reversible photoinduced changes observed in well-annealed amorphous films. This means that covalent (intramolecular) bond breaking is unlikely to be the dominant mechanism responsible for reversible photoinduced effects in amorphous chalcogenides.

The most significant photostructural change observed using Raman scattering was in the region of the low-frequency or boson peak at a wavelength of $25\,\text{cm}^{-1}$. This Raman peak is accounted for as signifying certain medium-range order in the amorphous structure [18–20]. The position and the intensity of the peak can be used to estimate the structural correlation length $R_c \approx (\omega_{max}/V)^{-1}$, where ω_{max} is the frequency of the peak maximum and V the velocity of sound, and to characterise the degree of structural ordering. The structural correlation length is generally associated with the dimensions of some cluster-like structure. For a-As_2S_3, the R_c value is $\sim 7.6\,\text{Å}$. The clusters may exist in the form of linked pyramids. In spite of the absence of definite conclusions on the nature of low-energy excitations in disordered solids, it is obvious that differences between the boson peak in the annealed and photodarkened states are due to changes in medium-range order. A decrease in the peak height and a shift to higher ω values may be attributed to an increase in structural randomness. This is consistent with an increase in structural disorder under illumination detected by EXAFS spectra [13, 21]. In spite of our experimental results on Raman scattering, PD is a photostructural change and is the result of changes in medium-range order (i.e. weakly linked AsS_3 pyramids move with respect to each other, as proposed in [22]). This mechanism does not involve the breaking of covalent bonds. At the same time, some authors (see the corresponding chapters of books and reviews [22–26]) suggest, on the basis of data obtained by indirect probes of local atomic structure, modification of short-range order accompanying reversible photoinduced changes. Clearly, unique, diffraction-like experiments are needed; probably they are the only ones that can give unambiguous evidence on the changes in short- or medium-range order accompanying reversible photoinduced changes. Although the former often is believed to be answered, Raman data presented in this chapter show that the main problems remain unresolved.

2.3 Summary

In this chapter the influence of thermal evaporation conditions and subsequent light and heat treatment on the atomic structure was studied. The major experimental findings of these studies may be summarised as follows.

The pronounced irreversible change observed in the Raman spectra of slowly deposited films on illumination is direct evidence for gross structural changes caused by optical irradiation. Such behaviour has been interpreted in terms of a photopolymerisation mechanism.

The films were found to become sulphur deficient as the deposition (evaporation) rate increased, because of partial fragmentation of As_2S_3 into As_3S_2 and sulfur during the deposition. Fast-deposited films are unstable and phase separation occurs upon light irradiation.

The reversible photoinduced effects in well-annealed amorphous films involve structural modifications in the medium-range order.

References

1. A. Madan, M.P. Shaw, *The Physics and Applications of Amorphous Semiconductors* (Academic, Boston, MA, 1988)
2. A. Feltz, *Amorphous Inorganic Materials and Glasses* (VCH, Weinheim, Germany, 1993)
3. S.O. Kasap, in *Handbook of Imaging Materials*, 2nd edn., ed. by A.S. Diamond, D.S. Weiss (Marcel Dekker, New York, 2002) p. 329, and references therein
4. K. Tanaka, in *Encyclopedia of Materials* (Elsevier, Oxford, 2001) p. 1123
5. S.A. Solin, G.N. Papatheodorou, Phys. Rev. B **15**, 2084 (1977)
6. R.J. Nemanich, G.A.N. Connell, T.N. Hayes, R.A. Street, Phys. Rev. B **12**, 6900 (1978)
7. V.I. Mikla, Yu.M. Vysochanskij, A.A. Kikineshi, et al., Sov. Phys. J. **11**, 73 (1983)
8. M. Frumar, A.P. Firth, P.J.S. Owen, Philos. Mag. B **50**, 453 (1984)
9. M. Frumar, Z. Polak, Z. Cernosek, J. Non-Cryst. Solids **256/257**, 105 (1999) 105
10. M.L. Slade, R. Zallen, Solid State Commun. **30**, 387 (1975)
11. P.J.S. Even, M.L. Sik, A.E. Owen, Solid State Commun. **33**, 1067 (1980)
12. U. Strom, T.P. Martin, Solid State Commun. **29**, 527 (1979)
13. C.Y. Yang, D.E. Sayers, M.A. Paesler, Phys. Rev. B **36**, 8122 (1987)
14. A. Bertoluzza, C. Fagnano, P. Monti, A. Semerano, J. Non-Cryst. Solids **29**, 49 (1978)
15. I. Zitkovsky, P. Boolchand, J. Non-Cryst. Solids **114**, 70 (1989)
16. K. Tanaka, Appl. Phys. Lett. **26**, 243 (1975)
17. A.E. Owen, A.P. Firth, P.J.S. Ewen, Philos. Mag. B **52**, 347 (1985)
18. A.J. Martin, W. Brenig, Phys. Stat. Solidi. B **63**, 163 (1974)
19. R.J. Nemanich, Phys. Rev. B **16**, 1655 (1977)
20. V.K. Malinovsky, A.P. Sokolov, Solid State Commun. **57**, 757 (1986)
21. L. Gladen, S.R. Elliott, G.N. Greaves, J. Non-Cryst. Solids **106**, 189 (1988)
22. G. Pfeiffer, M.A. Paesler, S.C. Agarwal, J. Non-Cryst. Solids **130**, 111 (1991)
23. G. Lucovsky, M. Popescu (eds.), *Non-Crystalline Materials for Optoelectronics* (INOE, Bucharest, 2004)
24. K. Tanaka, in *Handbook on Advanced Electronic and Photonic Materials and Devices*, ed. by H.S. Nalwa (Academic, San Diego, CA, 2001)
25. V.M. Lyubin, M.L. Klebanov, *Photo-Induced Metastability in Amorphous Semiconductors*, Chap.6 (Wiley, Weinheim, 2003)
26. M.A. Popescu, *Non-Crystalline Chalcogenides* (Kluwer, Dordrecht, The Netherlands, 2000)

Chapter 3
Optical Absorption and Structural Transformations in Arsenic Selenide Films

Optical absorption has been studied as a function of composition and temperature for poorly investigated Se-rich $As_x Se_{1-x}$ noncrystalline alloys. The samples were thin amorphous films and bulk glasses. It was shown for the first time that the absorption above the fundamental (Urbach) edge follows the Tauc law. The only exception to this law is pure selenium, for which a linear dependence of absorption coefficient on photon energy holds. The Tauc law is valid over a substantial range of absorption.

Although several reviews have appeared on various properties and applications of chalcogenide glasses, there is no thorough study of local atomic structure and its modification for Se-rich amorphous $As_x Se_{1-x}$ (a-$As_x Se_{1-x}$). This chapter is concerned with this problem. Structural transformations are examined by Raman scattering measurements of amorphous Se-rich $As_x Se_{1-x}$ ($0 \leq x \leq 0.2$) alloys. It is found that the molecular structure of amorphous Se (a-Se) on the scale of medium-range order differs from the structure of most inorganic glasses and may be placed between three-dimensional (3D) network glasses and polymeric ones. Further experiments show the existence of successive phases in laser-induced glass–crystalline transition with pronounced threshold behaviour. By comparing peak width, peak location and Raman intensity in the range of bond modes, it is derived that the changes occur non-monotonically with increasing As content. The composition-induced changes of the spectra are explained by crosslinking of Se chains. Under laser irradiation, the changes in the optical transmission, holographic recording properties and Raman spectra of a-$As_x Se_{1-x}$ films with $0 < x \leq 0.2$ have been examined. The dependence of the transmissivity and diffraction efficiency on the irradiation energy density shows two qualitatively different regions. Below the energy density threshold E_{th} only small changes in the local structure of the system can be detected. In the low-energy region, transient changes in transmissivity are observed. This behaviour may be explained qualitatively by associating such changes with alternation of deep defect states. Above E_{th}, the changes were attributed to crystallisation transformation. The corresponding Raman spectra reveal transformation of the system from amorphous phase to the crystalline phase under laser irradiation.

In physics, absorption of electromagnetic radiation is the way by which the energy of a photon is taken up by matter, typically the electrons of an atom. Thus, the

V.I. Mikla and V.V. Mikla, *Metastable States in Amorphous Chalcogenide Semiconductors*, Springer Series in Materials Science 128, DOI 10.1007/978-3-642-02745-1_3, © Springer-Verlag Berlin Heidelberg 2010

electromagnetic energy is transformed to other forms of energy. Usually, the absorption of waves does not depend on their intensity (linear absorption). The analysis of optical absorption is one of the most useful tools for understanding the electronic structure of amorphous semiconductors. The electronic absorption edge of amorphous semiconductors consists [1,2] of three spectral curves. In the absorption region higher than $10^3\,\mathrm{cm}^{-1}$, the absorption spectrum $\alpha\,(h\nu)$ follows the functional dependence $(\alpha h\nu)^{1/2} \propto h\nu - E_g^{\,T}$, and the optical bandgap energy is defined conventionally using $E_g^{\,T}$, the so-called Tauc gap. For smaller absorption coefficients, a temperature-independent Urbach tail with the form $\alpha \propto \exp(h\nu/E_U)$ is observed, where E_U is the Urbach energy. In the spectral region $\alpha \leq 100\,\mathrm{cm}^{-1}$, a weak absorption tail with a dependence $\alpha \propto \exp(h\nu/E_w)$ is observed, where $E_w \approx 300\,\mathrm{meV}$ (in As_2S_3) [1–3]. This absorption tail is influenced by impurities, but even in purified samples the tail still exists. It may be due to intrinsic mid-gap states. Interpretation of these spectral dependences still remains ambiguous.

Amorphous chalcogenides exhibit various structural changes under the action of light of photon energy comparable with their bandgap. These structural transformation cause changes preferentially in optical properties, but other physical properties are also "sensitive" to illumination. Various structural transformations caused by bandgap illumination are given in this section.

Considerable effort has been made by different research groups (see systematised data in [1–7] and references cited therein) to study the Urbach edge in As–Se amorphous semiconductors. These experiments show the unusual temperature and compositional behaviour of this fundamental property in the As–Se system. Such behaviour has been interpreted by specificity of microscopic structure and increasing disorder with decreasing temperature.

The universal occurrence of an Urbach edge in noncrystalline semiconductors, as is accepted in general, is preferentially caused by disorder, structural and thermal. The reader may find theoretical calculations indicating that random electric fields or a Gaussian site energy distribution both lead (Mott and Davis) to the exponential dependence of absorption on photon energy.

It is necessary to note that the absorption region above the Urbach tail has historically received less attention and was studied poorly. This fact is surprising because of the universal nature of the empirical Tauc law in amorphous solids. In the following we report a study of the Tauc region as a function of composition in Se-rich $As_x Se_{1-x}$ glasses.

The optical absorption was determined from the transmission measurements

$$T = \frac{(1-R)^2 \exp(-\alpha d)}{1 - R^2 \exp(-2\alpha d)}. \tag{3.1}$$

Here T is the transmission, R the reflectance, α the absorption coefficient and d the sample thickness. The value of R was taken corresponding to refractive index $n = 3.2$, in accordance with the compositions and photon energies typical of the present measurements.

It should be noted here that a-Se is a well-known exception to the Tauc law

$$\alpha h\nu = C\,(h\nu - E_{\mathrm{T}})^2 \tag{3.2}$$

and follows a linear law

$$\alpha h\nu = C_1(h\nu - E_1)\,. \tag{3.3}$$

The results obtained for Se amorphous films and bulk samples at room temperature are in full agreement with that given in [7]. It was found that the Tauc law may be applied to glassy samples of composition $As_x Se_{1-x}$ from 1 at% to 40 at%. This is a general tendency of optical absorption in the above mentioned compositional range. Pure a-Se samples obey the linear law from slightly above the extrapolated gap to the highest measurable energies. The addition of less than 1 at% As causes a transition from the anomalous linear behaviour to the universal Tauc law.

In recent years there has been enhanced interest in structural studies of chalcogenide glasses owing to their actual and potential use in electrical and optical devices. As-containing chalcogenides, especially Se-rich alloys, have usually been studied in thin-film, bulk and fibre forms. Since a change in the structure can influence photogeneration, charge-carrier transport, trapping and other important fundamental properties, knowledge of the molecular structure of such materials is needed for further improvement of their characteristics.

Undoubtedly, elemental semiconductors are useful and suitable model objects for studying the influence of structure on physical properties. This is particularly true with respect to selenium. In the past, the latter has been successfully used in photocells, rectifier diodes and solar cells. In its amorphous form, selenium has a good application as a photoreceptor in copying machines and X-ray imaging plates [4,5]. It is necessary to note that this material differs from other amorphous chalcogenides in several respects:

- Selenium is the only one that vitrifies as an elemental glass and is fairly stable at room temperature
- Compositional (chemical) disorder is absent
- Charge carriers of both polarities are mobile at room temperature
- The spectral dependence of the optical absorption in the region above the Urbach tail is unusual
- There exists a so-called "non-photoconductive gap"
- And the gap state density in the mobility gap is relatively low compared to other amorphous materials

Despite the increasing commercial using of a-Se in various applications (see [3] and references therein), for example as a promising X-ray flat panel detector for medical purposes [4,5], its structure is not fully understood.

On the other hand, binary noncrystalline semiconductors of the As–Se system are also of continued scientific and practical interest because of real opportunities for their technological use (e.g. as functional elements of multilayer photoreceptors in xerography). Among them the stoichiometric composition $As_2 Se_3$ and compositions in the range 30–50 at% As are perhaps the most studied ones [2–4]. As for most stable binary glasses in the As–Se system, which are also the subject of this book, atomic ratios can be varied in a wide range. Although the information about the various physical properties of Se-rich alloys is not so extensive [2–11], their compositional dependence manifests extrema or thresholds also in the range 6–12 at% As. It is necessary to accentuate that the As–Se amorphous alloy system displays main extrema of various properties at the composition where the valence requirements appear to be satisfied, that is, at the stoichiometric composition. It seems to be reasonable to connect the mentioned non-monotonic behaviour with a specific character of local structure changes.

Another prominent feature of the materials studied consists in the following. Certain types of glasses, whose common feature is the presence of the chalcogen atoms sulfur, selenium, and tellurium, exhibit various photoinduced phenomena (the reader may refer to [12]). Among these are photostructural transformations and photocrystallisation phenomena: a change in optical, electrical and other physical properties is observed. The phase transformation of selenium and its alloys can also be induced relatively simply by laser illumination [2, 13–15]. Reasonably, this unique property makes them attractive for optical data storage and holographic recording. Many experimental results using selenium and its alloys have been reported [4], but few cases of phase transformation properties were mentioned.

Raman scattering is a very powerful experimental technique for providing information on the constituent structural units in a given material [16]. In the present chapter, Raman scattering in pure a-Se and Se-rich As–Se amorphous films is studied. Below we attempt to clarify the structural transformations induced by light treatment and compositional changes. We focus our attention mainly on photocrystallisation transformations, although photostructural changes are also considered. In addition, the composition-induced structural modifications in a-$As_x Se_{1-x}$ are also analyzed. As the detailed analysis of Raman data shows, some discontinuity of atomic arrangement with rising As content exists.

3.1   Sample Preparation and Measurement Technique

The samples used in these studies were amorphous films, about $10\,\mu$m thick, prepared by vacuum thermal evaporation of the powdered $As_x Se_{1-x}$ melt-quenched material at a rate of $1\,\mu$m min^{-1} onto quartz substrates held at room temperature as well as polished mirror-like parallelepipeds of vitreous $As_x Se_{1-x}$. The $As_x Se_{1-x}$ bulk glasses were prepared according to the conventional melt-quenching method. Annealing of the films was carried out in air at ambient pressure and at temperatures below the glass transition temperature. Thin-film samples were kept in complete darkness until measured to minimise exposure to light sources, which could lead to changes in the properties and structure of the films. It is important accentuate that after annealing their Raman spectra become indistinguishable from the corresponding spectra of melt-quenched glassy samples. It should be noted that for Se-rich compositions studied the difference in the spectra of melt-quenched and as-deposited samples is not so "dramatic" as in the case of the As–S system, which is considered in Chap. 2, and as reported also by Nemanich et al. [17] (see also [18]).

Right-angle Raman spectra were measured using a RAMANOR U-1000 spectrometer. The spectral slit width was ca. $1\,cm^{-1}$ and the excitation wavelength 676 nm. Raman spectra of the amorphous films were recorded with sufficiently low incident laser-beam power densities, $P = 3$–$5\,mW$, to avoid photostructural changes. The latter is known to transform the Raman spectra in a manner similar to that reported in [19] for a-Se. The identity of the experimental spectra obtained from different points of the sample and the good reproducibility of the spectra in repeated scans (the time required to scan one spectrum in the spectral range 100–300 cm^{-1} is about 5 min) show that photodarkening (PD) did not play a role in the subsequent Raman measurements.

In the transmission PD experiments, the samples were illuminated at near-normal incidence by argon–ion and helium–neon lasers, operating, respectively, at 514 nm and 633 nm. These wavelengths were convenient because they correspond to near maximum photoresponse. The transmission of the sample was probed using a portion of the He–Ne output and detected by a photomultiplier. The intensity of the probe beam was kept low ($P \leq 0.01\,W\,cm^{-2}$) and had no measurable effect on the sample, in either producing or erasing the PD. The sample transmission was measured at 633 nm, during and after PD.

The structure of photocrystallised films was investigated using X-ray diffraction (XRD).

The holographic gratings were recorded on thin-film samples of $As_x Se_{1-x}$. The present grating technique is the conventional method, in which a grating having a pitch $\Lambda = \lambda/\left(2\sin\left(\Theta/2\right)\right)$ is produced by two interfering beams of wavelength λ (633 nm) intersecting at a sample surface with an angle Θ (ca. 40°). The diffraction efficiency $\eta = I/I_0$, where I_0 and I are the corresponding intensities of the reading beam and of the first-order diffracted beam, is measured as a function of the exposure time.

3.2 Raman Scattering in Amorphous Selenium

In Fig. 3.1, a typical Raman spectrum of a-Se measured at low incident radiation power density, ca. 3 mW, is shown. The stable level of the scattered light intensity and the good reproducibility of the spectrum in the repetitive cycles clearly indicate the absence of any structural changes in a-Se induced by laser irradiation of such power density. On the high-frequency side ($\omega = 100$–$300\,cm^{-1}$) the spectrum contains an intense peak at 255 cm^{-1} and a feature (shoulder) at 237 cm^{-1}.

The above features are in good agreement with previously reported data [20–23]. In the low-frequency region one can observe a broad peak with ω_{max} at 16–20 cm^{-1}. This so-called boson peak occurs in the low-frequency region of the Raman spectra of all amorphous and vitreous solids. An analysis of the spectral form of the boson peak is given in Fig. 3.2.

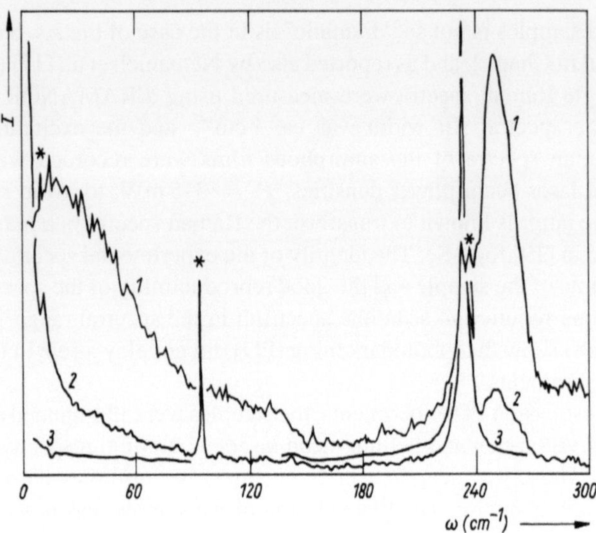

Fig. 3.1 Raman spectra of (1) amorphous and (2), (3) photocrystallised selenium. Experimental details: (2) after 5 min exposure of the sample to 10 mW, (3) after 30 min exposure to 10 mW. *Asterisks* indicate laser plasma lines

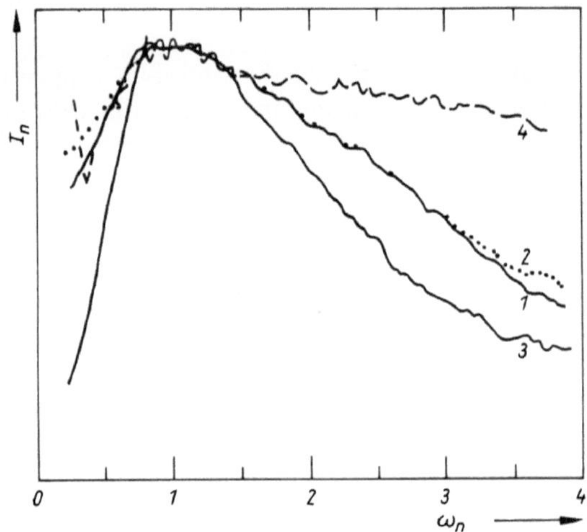

Fig. 3.2 Low-frequency Raman spectra of different glasses on a scale $\omega_n = \omega/\omega_{max}$. (1) Se ($\omega_{max} = 17\,\mathrm{cm}^{-1}$ at $T = 100\,\mathrm{K}$), (2) Se ($\omega_{max} = 17\,\mathrm{cm}^{-1}$ at $T = 300\,\mathrm{K}$), (3) As$_2$Se$_3$ ($\omega_{max} = 26\,\mathrm{cm}^{-1}$, $T = 300\,\mathrm{K}$ [24]), (4) poly(methylmethacrylate) (PMMA) ($\omega_{max} = 17\,\mathrm{cm}^{-1}$ [25]); see also [26–30] for the structural characterisation of amorphous chalcogenides

Normalised Raman spectra $I_n = I/[\omega(n(\omega)+1)]$ for a-Se and a series of samples of other compositions are given on the same energy scale $\omega_n = \omega/\omega_{max}$, where $n(\omega) + 1 = 1/(\exp(\hbar\,\omega/kT) - 1) + 1$ is the boson factor for the Stokes component.

When we try to identify certain vibration bands observed in the Raman spectrum of a-Se, some difficulties arise. Initially it was proposed to interpret the a-Se Raman spectrum in analogy with sulfur – on the basis of a molecular approach. That is, the main vibration band was considered to be the superposition of the peaks at 237 and 255 cm^{-1} characteristic of chains and rings, respectively [21]. However, further experimental data have caused some doubts to be cast. In such a case one would expect to observe a discernible difference in the contributions of the 237 and 255 cm^{-1} modes to the main vibrational band in samples prepared under different conditions (e.g. substrate temperature during deposition for amorphous films or quenching rate for glassy samples). This may be caused by changes in the rings-to-chains ratio. Therefore, it is clear that the spectral region 200–300 cm^{-1} is unsuitable for ring diagnostics. This is consistent with the conclusions of [22].

The Se$_8$ peak (112 cm^{-1}) could not be detected in the present experiment. Probably, this is connected with its weakness. This fact evidently indicates the low concentration of rings. Consequently, one can associate the spectral features in the main vibration band, the 255 cm^{-1} peak and the shoulder at 237 cm^{-1}, mainly with chain vibrations.

The low-frequency region $0 < \omega < 100$ cm^{-1}, in which the boson peak appears, is of special interest. Note that such a low-frequency peak seems to be a universal feature of all amorphous materials [24, 31]. It has been found that the spectral form of the boson peak is the same for a wide range of oxides, chalcogenides, and low-molecular-weight organic glasses, and coincides with that of As$_2$S$_3$ bulk glass [16, 17, 24]. The universal form of the low-frequency peak is due to the universality of glassy material on the scale of medium-range order $L \approx V/\omega_{max}$ of 1–2 nm (V is the sound velocity) [31]. For the case of a-Se, it is observed (Fig. 3.2) that the spectral form of the boson peak essentially differs from that characteristic of the majority of inorganic glasses. The spectral form of the boson peak in a-Se seems to be intermediate between that in polymeric and low-molecular-weight glasses.

This result can be explained by a preferentially chain-like structure of a-Se. The latter may form a structure similar to the structure of the linear polymer PMMA. In other words, with regard to its structure on the scale of medium-range order, a-Se may be placed between 3D network glasses and polymeric ones. Examples of intermediate-range order in elemental and compound materials, including a-Se, have been extensively discussed in [24, 31–33].

There exists another possible explanation. Se has its glass transition temperature near room temperature. Therefore it is in a well-annealed state. It is known [31] that the intensity of the boson peak relative to the main bond modes in the Raman spectrum significantly decreases as the structural order of the sample increases (e.g. in "equilibrated" or annealed samples). This decrease may lead, in turn, to an increased contribution from other modes to the high-frequency side ($\omega \geq \omega_{max}$) of the boson peak. This is clearly seen if we compare the shape of corresponding peaks in a-Se and As$_2$S$_3$ (Fig. 3.2).

3.3 Composition Dependence of Raman Bands in Amorphous Se-Rich $As_x Se_{1-x}$ Alloys

In this section, experimental results on Raman scattering spectra for Se-rich amorphous semiconductors $As_x Se_{1-x}$ are discussed. In Fig. 3.3 typical Raman spectra of a-Se and As–Se alloys with As content up to 5 at% are shown. The major spectral feature in the high-frequency region is the $255\,cm^{-1}$ band. Another prominent spectral feature, which is not shown in this figure, is the broad peak with ω_{max} at 16–$20\,cm^{-1}$. As mentioned, the latter is characteristic for Raman scattering of all amorphous solids and glasses.

In the following we consider only the high-frequency region. The weak feature observed at $112\,cm^{-1}$ in the a-Se spectrum diminishes with As addition and at 5 at% completely disappears (see inset in Fig. 3.3). At the same time the difference in the spectra in the region of the main vibration band is obvious. Thus, with increasing As content the transformation of the Raman spectrum in this region is retraced (Fig. 3.4).

The most important points are the following.

- Spectrum broadening with increasing As addition.
- Growth of scattered light intensity from the low-frequency side of the main maximum ($255\,cm^{-1}$).
- Appearance of a broadened band at 220–$230\,cm^{-1}$ (this band is most intense in the Raman spectrum of $As_{0.4}Se_{0.6}$). It should be noted also that the main maximum is slightly shifted to higher frequency for a-$As_x Se_{1-x}$ with respect to Se ($255\,cm^{-1}$).

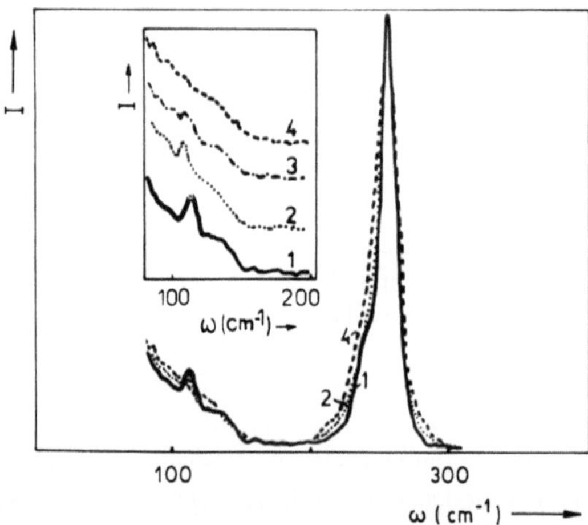

Fig. 3.3 Comparison of the Raman spectra of a-Se (1) and $As_x Se_{1-x}$: $x = 0.02$ (2), 0.04 (3), and 0.05 at% (4). Each trace has been normalised to the same peak ($255\,cm^{-1}$) intensity. The *inset* shows the bending mode region

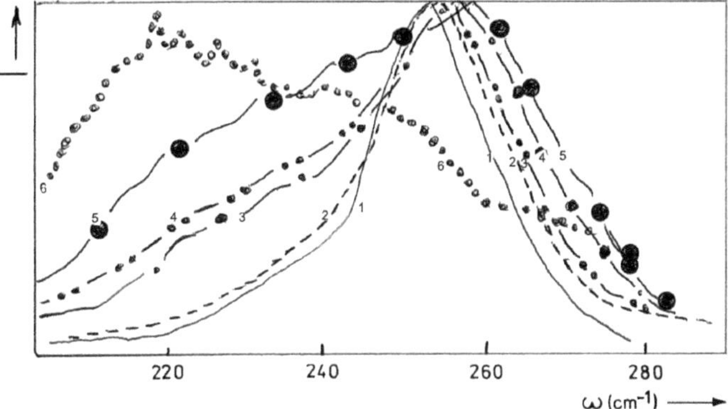

Fig. 3.4 Raman spectra of a-As$_x$Se$_{1-x}$ films: $x = 0$ (1, *solid line*), 0.05 (2, *dashed line*), 0.10 (3, *dashed-dotted line*), 0.12 (4, *dashed–double dotted line*), 0.20 (5, *dashes and large circles*), and 0.40 (6, *large dots*). We have used unconventional notation for the Raman spectra to make them distinguishable

- The intensity of the 220–230 cm^{-1} band in the As concentration interval 0–5 at% remains practically unchanged. Then, at 6 at% As, an increase of the band intensity occurs. A gradual intensity rise is observed for the band at 220 cm^{-1} as the As content is further increased. For As content exceeding 35 at% the band dominates the Raman spectrum.

It seems to be reasonable to approximate the observed Raman spectra of As$_x$Se$_{1-x}$ as a superposition of the spectra of a-Se and As$_{0.4}$Se$_{0.6}$. The corresponding calculations have been performed. These calculations yield that a systematic discrepancy between approximated and experimental spectra is observed. For the latter, greater values of the main peak width are typical.

Figure 3.5 shows difference spectra obtained by subtracting the As$_{0.4}$Se$_{0.6}$ spectrum from experimental Raman spectra. The relative contribution of the As$_{0.4}$Se$_{0.6}$ spectrum was fitted to the region around 230 cm^{-1}, where the contribution from pure Se was negligibly small. It is obvious that after such a procedure some peak remains, the width and position of which differs from that for a-Se. Based on the data given in Fig. 3.5, values of the parameter A were estimated. This parameter represents the ratio of the integrated Raman intensity in the interval limited by the typical frequencies of AsSe$_{3/2}$ unit vibrations (205–230 cm^{-1}) to the integrated intensity of the whole spectrum of valence vibrations

$$A = \int_{205\,\text{cm}^{-1}}^{230\,\text{cm}^{-1}} I\,d\omega \bigg/ \int_{205\,\text{cm}^{-1}}^{290\,\text{cm}^{-1}} I\,d\omega, \tag{3.4}$$

(here I is the intensity of the corresponding Raman band). Figure 3.6 shows that the dependence of A on x, $A = f(x)$, is not smooth: parameter A increases rapidly around 6 at% As. The change of scattered intensity with composition for

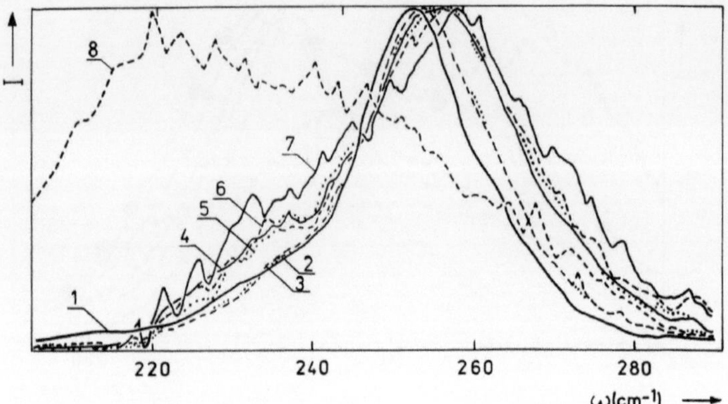

Fig. 3.5 The difference spectra (for details see the text) for amorphous As_xSe_{1-x}. $x = 0.04$, 0.05, 0.06, 0.08, 0.10 and 0.20 at% for curves 2–7, respectively. For comparison, the Raman spectra of a-Se (1) and $As_{0.4}Se_{0.6}$ (8) are also shown

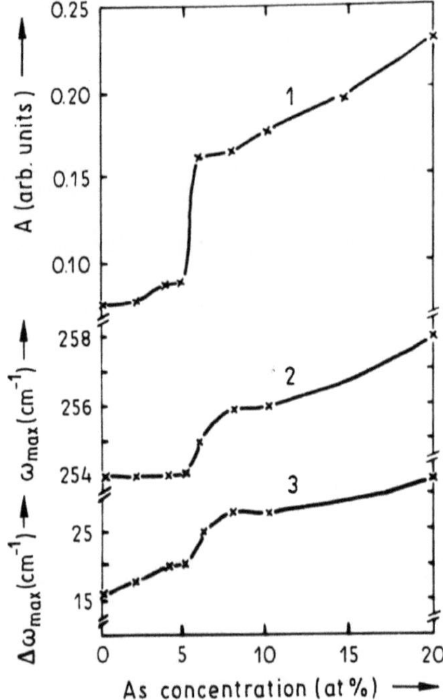

Fig. 3.6 Dependence of the parameters (1) A, (2) ω_{max} and (3) $\Delta\omega_{max}$ on composition

the frequency range 240–270 cm^{-1} has a smoother character. In the same figure, the dependence of the peak frequency ω_{max} and its width $\Delta\omega_{max}$ on As content for the corresponding spectra is displayed. It is important to note the similarity of the composition dependence of A, ω_{max} and $\Delta\omega_{max}$.

An attempt to simulate As_xSe_{1-x} Raman spectra by a superposition of two constant spectral forms, one belonging to a-Se, the other to $As_{0.4}Se_{0.6}$, failed. Onari and coworkers [34, 35] were the first to use such an approach. On the other hand, the experimental spectra could be approximated assuming a considerable broadening of chain vibrations and their frequency displacements. We consider that such an approach is correct and that the difference spectra themselves are convincing arguments in favour of it: the change of the Raman spectra with composition together with parameters A, ω_{max}, and $\Delta\omega_{max}$ (Figs. 3.5 and 3.6) support this suggestion.

Composition-dependent studies of the physical properties of binary and ternary chalcogenide glasses give evidence for the existence of mechanical and chemical thresholds at certain compositions of these materials [36–38]. The As–Se system displays main extrema of various properties at the stoichiometric composition (the mechanical and chemical thresholds coincide at $x = 0.40$). There seem to exist (see experimental results published by Wagner and Kasap [39] and data presented here) an additional threshold at $0.06 \leq x \leq 0.12$. It can be argued that the non-smooth behaviour observed in the concentration dependence of glass transition temperature, density, etc. [6] in this range originates from changes in bond topology [37, 40]. We assume that in Se-rich glasses the network is dominated by Se atom chains (quasi-1D network) and addition of As atoms lead to branching owing to three-fold coordination of As atoms. Some publications [39, 41, 42] provide, we believe, a new approach to the problem of local bonding in amorphous chalcogenides. The anomalous behaviour near $x \approx 0.06$ is ascribed to the disappearance of Se_8-like segments. From the point of view of configuration, it is suggested that the number of *cis* configurations starts to decrease, so that intermediate-range correlation is modified. The considerable reduction in the vibration mode at $\sim$112 cm^{-1}, which is associated with cis segments, strongly supports this suggestion. Changes in the Raman spectrum with composition allow us to conclude that incorporation of As leads to crosslinks between chain-like or ring-like segments of a-Se.

There are strong indications that the compositional dependence of physical and chemical properties has no connection with chemical ordering. In fact, the binary As_xSe_{1-x} alloys exhibit extrema in compositional dependence of the density not only at the $As_{0.4}Se_{0.6}$ composition, but also for the nonstoichiometric chalcogen-rich $As_{0.06}Se_{0.94}$ and also for pnictogen-rich $As_{0.6}Se_{0.4}$ samples. This means that the x dependence of the density originates from changes in bonding topology.

3.4 Laser-Induced Structural Transformation of As_xSe_{1-x} Amorphous Films

The PD phenomenon continues to attract extensive interest, since this is a simple, at first sight, bulk phenomenon that is characteristic of chalcogenide glasses. That is, it does not appear in the corresponding crystal. Although PD and related changes have been extensively studied, the mechanism of this unique phenomenon is not yet elucidated. Figure 3.7 illustrates the development of the PD effect (i.e. the change

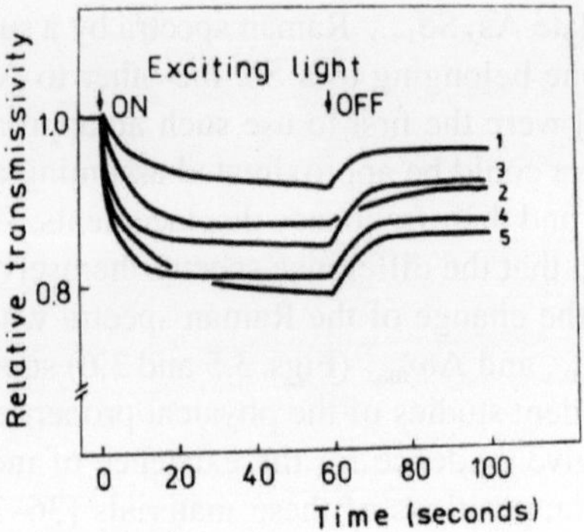

Fig. 3.7 Time dependence of photoinduced changes in transmissivity in As_xSe_{1-x} films. $x = 0$, 0.02, 0.05, 0.08 and 0.12 for curves 1–5, respectively

in transmissivity) at room temperature as a function of time with the illumination turned on and off.

During on-periods, the intensity of illumination remained constant. All As_xSe_{1-x} ($0 \leq x \leq 0.2$) films show a decrease in relative transmissivity T_{rel} (i.e. the ratio between the transmissivity T_{ir} of irradiated As_xSe_{1-x} films and the transmissivity T_u of untreated films) with illumination. On illumination there is an initial decrease with illumination time up to the constant value. For this rapid change to be observed, an exposure of $10\,J\,cm^{-2}$ is required. The degree of change increases with As content. When the illumination is interrupted, the transmissivity increases (bleaching). The magnitude of bleaching decreases with addition of As. Note that a-Se and As_xSe_{1-x} ($0 < x < 0.20$) exhibit appreciable self-annealing effects if photostructural changes are induced near room temperature [12, 29, 40, 43–45]. The reason for this is that the glass transition temperatures ($T_g = 41$ and 57°C for Se and $As_{0.05}Se_{0.95}$, respectively [6]) are close to the illumination temperature. A similar transient transmission change was observed for all samples, irrespective of their composition or previous treatment, such as illumination or annealing in dark. These changes in transmissivity in cycles of on–off illumination can be repeated reproducibly many times. In that sense we consider them reversible. These reversible changes appear to saturate after $\sim100\,J\,cm^{-2}$ integrated exposures at exciting light intensities $<0.5\,W\,cm^{-2}$.

Increasing exposure (in our case for intensities $>0.6\,W\,cm^{-2}$) caused a significant irreversible change in transmissivity, which we attribute to a crystallisation transformation as discussed below.

After an initial response corresponding to the transient transmission change, there is a slowly varying response. The results, reflecting this situation, are shown in Fig. 3.8. The slowly varying portion of the darkening curve is followed by a decrease in transmission. The latter appear to indicate actual crystallisation (see the corresponding Raman data). The transmission change completed by a saturation region. The latter is intensity dependent.

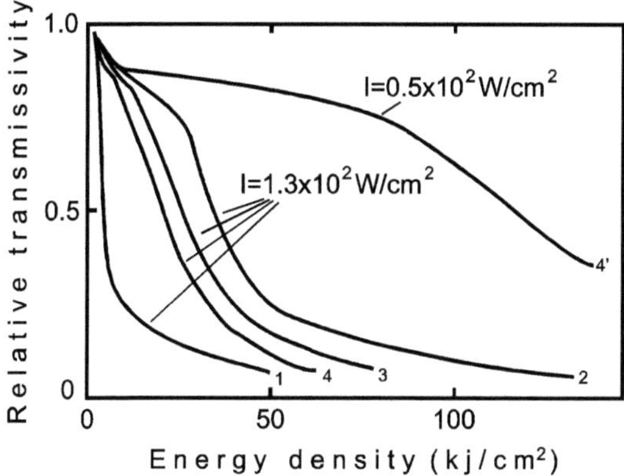

Fig. 3.8 Evolution of the relative transmissivity change in As_xSe_{1-x} samples exposed to laser illumination at 633 nm. Thickness is $1.2\,\mu m$ and intensity $1.3 \times 10^2\,W\,cm^{-2}$. $x = 0, 0.05, 0.08$ and 0.12 (curves 1–4, respectively). For a-$As_{0.12}Se_{0.88}$, the corresponding curve is also shown for lower intensity ($0.5 \times 10^2\,W\,cm^{-2}$)

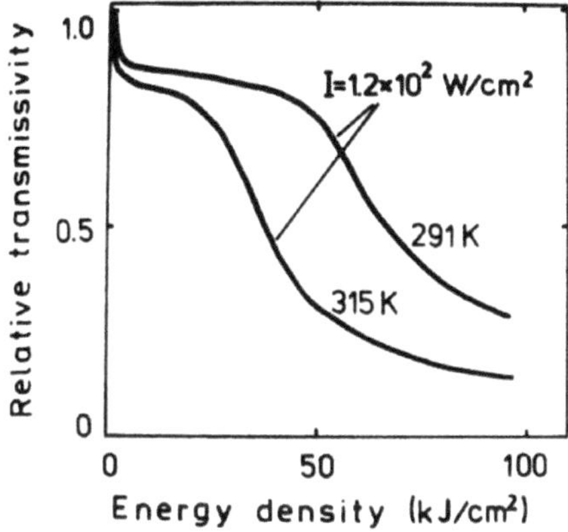

Fig. 3.9 Change in the relative transmissivity in $As_{0.05}Se_{0.95}$ samples induced by laser irradiation at 291 and 315 K. Inducing intensity was $1.2 \times 10^2\,W\,cm^{-2}$

The total change in transmission may attain $\sim$90% at such an exposure. The addition of arsenic to a-Se and/or decreasing the temperature (Fig. 3.9) delays the crystallisation onset, which starts at higher exposure times. It is necessary to note that during laser exposure oxidation of films may take place. In Raman spectra of As_xSe_{1-x} samples only bands characteristic of this composition are present. We believe that this is a strong argument that oxidation may be ruled out in this case.

When a sample of a-Se is exposed to the 490 nm light ($I \approx 1.9\,W\,cm^{-2}$), the transmission characteristics are similar to those induced by 633 nm light. This similarity is the case for all compositions examined. The observed transient transmission change is $\sim$8%. At the same time we cannot induce crystallisation-related effects even for a total exposure of $10^4\,J\,cm^{-2}$ at $10^2\,W\,cm^{-2}$.

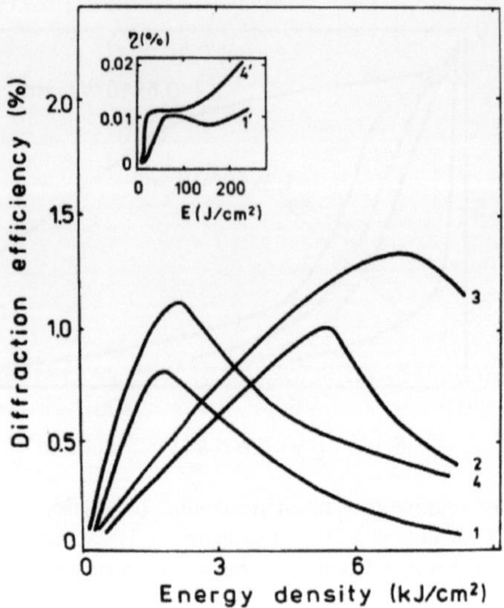

Fig. 3.10 Diffraction efficiency η as a function of energy density for a-As$_x$Se$_{1-x}$. $x = 0$, 0.05, 0.08 and 0.12 (curves 1–4, respectively). The *inset* shows the low-energy region for a-Se (1′) and a-As$_{0.12}$Se$_{0.88}$ (4′). $I = 0.6\,\mathrm{W\,cm}^{-2}$

Holographic gratings were recorded on Se and As$_x$Se$_{1-x}$ thin films. Figure 3.10 shows a typical result for light intensities diffracted from gratings formed on As$_x$Se$_{1-x}$ film. The low-energy region we attributed to transient effects, while the high-energy region is reasonably connected to photocrystallisation-related permanent changes.

The tail observed after η reached its maximum at higher exposure is due to the fact that photocrystallisation increases the optical absorption for the probing light.

The light-scattering spectra recorded after irradiation with an intense laser beam (the exposure conditions were chosen to be identical to those that cause the above described transmission change) reveal well-defined changes in the structural organisation of the films.

We may summarise the general features of the observed transformation of Raman spectra in the range of bond bands for a-Se. (1) There is (Fig. 3.11) a certain threshold energy E_{th} of the incident radiation. (2) Below E_{th}, no changes of Raman spectrum are observed. Here we note that the only exception is the so-called boson peak detected at around $17\,\mathrm{cm}^{-1}$, which is weakened by illumination. (3) Above E_{th}, an increase of the incident energy density modifies the spectra.

The present experimental results permit us to distinguish three successive stages of photocrystallisation in a-Se with regard to irradiation energy density. First of all it is necessary to point out the absence of any significant structural transformations in films and bulk samples at $E_{th} \leq 4\,\mathrm{J\,cm}^{-2}$. This is strongly supported by the identity of the Raman spectra recorded in repetitive cycles. At the first stage, which is induced by irradiation with incident energy density $\geq 3.8\,\mathrm{J\,cm}^{-2}$, microcrystallite

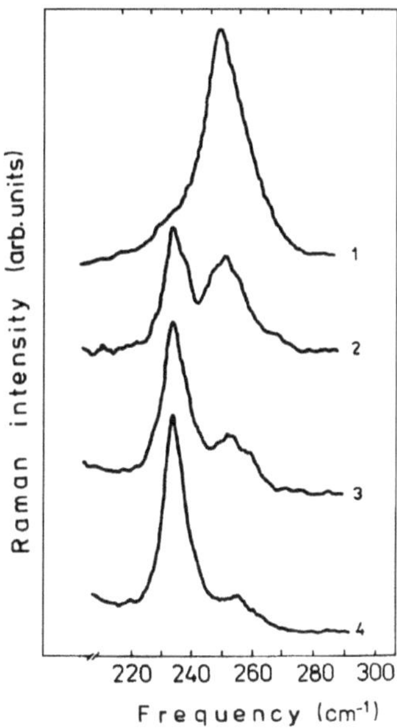

Fig. 3.11 Laser-induced Raman spectra transformation for a-Se. (1) Reference spectrum of amorphous state. (2–4) Spectra after exposure to $E = 3.8$, 11.4 and 19×10^3 J cm^{-2}, respectively. $I = 1.3 \times 10^2$ W cm^{-2}

formation probably takes place. In such a case the 255 cm^{-1} peak dominates the Raman spectra.

The second stage of photocrystallisation ($\sim$11 J cm^{-2}) is characterised by an enlargement of microcrystalline units. This is demonstrated by the growth of the 237 cm^{-1} peak. Finally, at $\sim$20 J cm^{-2}, photocrystallisation practically takes place. This stage is marked by a dramatic increase and narrowing of the peak at 237 cm^{-1} with irradiation time. At the same time, the 255 cm^{-1} peak becomes more and more suppressed with respect to other Raman-active modes, and finally it completely disappears. At this last stage of photocrystallisation the absence of the low-frequency (17 cm^{-1}) peak is also characteristic.

It seems to be reasonable to assume a thermal mechanism for the observed structural transformation, caused by laser heating of the sample. This suggestion is strongly confirmed by the clearly manifested threshold behaviour. Additional support comes from the fact that at low temperature the threshold power (e.g. 20 J cm^{-2} at $T = 100$ K for a-Se) that is necessary for changes in Raman spectra to be observable exceeds several times that for changes at $T = 300$ K. Note that the changes under examination differ qualitatively from the well-known PD phenomenon. The latter takes place at any value of irradiation power; threshold behaviour is not characteristic of it. The magnitude of PD depends mainly on the amount of absorbed energy and significantly increases with temperature lowering.

Table 3.1 E_{th} values as a function of As concentration in $As_x Se_{1-x}$ alloys

As concentration (at%)	0	5	8	12	20
E_{th} (kJ cm^{-2})[a]	2 ± 1	12 ± 3	7 ± 2	8 ± 2	45 ± 5

[a]The E_{th} values are given for the case $I = 1.3 \times 10^2 \, W \, cm^{-2}$

The greater efficiency in films in comparison with bulk samples is an established feature of PD. In contrast, in our case, the threshold power densities for a-Se films are found to be higher than for bulk samples. According to some results [29, 30], discernible, reversible PD in a-Se at $T = 100$ K occurs at photon energy $h\nu \geq 2.0$ eV with maximum efficiency at ~ 2.4 eV. The exciting irradiation energy $h\nu = 1.84$ eV seems to be too low to induce significant PD at $T = 100$ K. At the same time the probability of transient PD effects relaxing when the irradiation stops at $T = 300$ K for $E \approx 3.8 \, J \, cm^{-2}$ cannot be definitively ruled out.

With regard to $As_x Se_{1-x}$ Raman data, the principal results are the following:

- The spectra of the $As_x Se_{1-x}$ amorphous alloy samples before irradiation were the same as those reported in [8, 17, 20, 22, 23, 34, 35, 45].
- The value of E_{th} necessary for changes in Raman spectra to be observable vary with addition of As (see Table 3.1).
- Under irradiation with $E > E_{th}$ the recorded spectra clearly show a narrowing and increase in the intensity of the 237 cm^{-1} Raman band.
- In $As_x Se_{1-x}$ samples with $x \leq 0.05$, no additional (to that recorded for pure Se) photoinduced changes in the Raman spectra are observed (see Fig. 3.12 and compare it with the results for a-Se). It should be accentuated that, on introducing such a relatively large quantity of As additives, there is no appreciable influence on the photocrystallisation product. It is well known that Se is likely to be photocrystallised at ~ 350 K [6, 13]. Reasonably, XRD patterns of the samples have been measured. Since the illumination region is of ~ 3 mm in radius, the pattern is noisy. However, we can clearly see four crystalline peaks located at $2\theta = 24°$, $30°$, $41°$ and $45°$. The peaks can be indexed, respectively, as 100, 101, 110 and 111 of the trigonal (hexagonal) Se crystal [46].
- For As content > 15 at%, the main result of this study is the appearance and disappearance of new Raman bands typical of $As_{0.4} Se_{0.6}$. For example, Fig. 3.12 shows the appearance of an additional peak (~ 264 cm^{-1}) superimposed on the amorphous peak (~ 255 cm^{-1}). With a further increase of the irradiation energy density, crystallisation starts immediately. Here we should note an evolution qualitatively similar to that shown for pure Se and $As_{0.05} Se_{0.95}$ in the shape of spectra over the concentration range 0–20 at%. In the final stage of photocrystallisation, the spectra of $As_x Se_{1-x}$ alloys are free of key crystalline features that occur [47] in the spectra of crystalline $As_2 Se_3$. It is of particular significance that only the 237 cm^{-1} band of trigonal selenium contributes to the spectra of photocrystallised $As_x Se_{1-x}$ ($0 < x < 20$) films.

In order to discuss further the changes in the optical transmissivity and diffraction efficiency with irradiation energy density, we will divide the T_{rel} and η versus E data

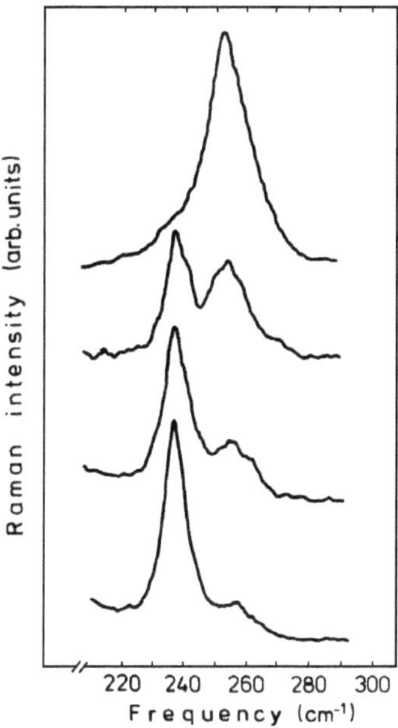

Fig. 3.12 Development of the photocrystallisation effect in $As_{0.05}Se_{0.95}$ as a function of exposure: (1) Reference spectrum of the amorphous state. (2,3) After exposure to 15 and $25 \times 10^3 \, \mathrm{J\,cm^{-2}}$, respectively. $I = 1.3 \times 10^2 \, \mathrm{W\,cm^{-2}}$

into two qualitatively different regions with a separation at E_{th}. We define E_{th} (e.g. $E_{th} = 2 \times 10^3 \, \mathrm{J\,cm^{-2}}$ at $I = 1.3 \times 10^2 \, \mathrm{W\,cm^{-2}}$ in the case of pure Se) as the energy density for which the system is not yet perturbed structurally (on the scale of short-range order) by laser irradiation. The absence of any significant bonding changes in films is supported by the identity of the Raman spectra recorded in repetitive cycles. This result holds for both pure a-Se and Se-rich As_xSe_{1-x} alloys. For energies less than E_{th} reversible PD and transient transmission changes are observed. These effects are characteristic of the amorphous phase and the system remains in the amorphous phase under or even after irradiation. The lack of any noticeable variation in the transient behaviour T_{rel} for samples of different substrate material and film thickness excludes the possibility of the effect being due to small changes in sample temperature during and after illumination, that is, to photoinduced heating.

We have recently reported similar dynamical photoinduced changes in some photoelectronic properties detected by time-of-flight and xerographic techniques [40, 48]. These experiments may provide the first evidence that deep defects can be altered temporarily by room-temperature irradiation. Note that there is a close correlation between the recovery of optical parameters and photoelectronic characteristics in exposed samples. Although a complete correlation of microscopic structural modifications with macroscopic PD phenomena must await further experimental measurements, it is only natural that we relate the transient changes in the transmissivity with changes in deep defect states. We identify such centres as arising from native (thermodynamic) structural defects (e.g. C_3^+ and C_1^- in a-Se). Bandgap light can probably initiate conversion of traps of small cross section to those of larger cross section [4, 33].

In contrast, above E_{th}, all the observed irreversible changes may be attributed to optical-constant variations and modifications in the kinetics of light-induced crystallisation. The present experimental results resolve successive stages of photocrystallisation in a-Se (these were mentioned above). In such a photocrystallisation process, a-Se undergoes a transformation to trigonal selenium, which is the most stable modification. Raman scattering studies together with XRD data gave an unambiguous indication of trigonal selenium.

On the basis of the present Raman data, we conclude that the features of the photocrystallisation effects in $As_x Se_{1-x}$ alloys with $x < 0.15$ are qualitatively the same as those in a-Se. Some deviation in pre-crystallisation behaviour of $As_{0.2}Se_{0.8}$, namely the appearance and disappearance of the weak quasi-crystalline peak at $264 \, cm^{-1}$, probably indicates $As_2 Se_3$-like cluster creation and annihilation. The latter could be clusters with a more ordered structure than that existing in the amorphous phase. At the same time, they are not yet microcrystallites with inherent Raman peaks. It seems to be reasonable that the environment of the clusters prevents their growth and transformation into microcrystallites. Other Se clusters reach the critical size required for microcrystallite formation. After that, sample exposure to $E > E_{th}$ crystallised selenium, while the As-containing clusters remained in the amorphous phase. The above explanation is in agreement with the results of a study of laser-induced structural transformations in glassy $As_2 Se_3$ [33] and also with the mechanism proposed by Phillips [49].

It is known [6] that As is an effective additive to decrease the tendency to crystallise. Our experimental results, namely the greater value of E_{th} for a-$As_x Se_{1-x}$ films, indicate that the addition of As effects the suppression of the crystal nucleation and growth in a-Se. The long Se chains in a-Se branch at the site of As atoms. The length of Se chains becomes short [35, 45] and the a-Se cannot easily crystallise with increasing As concentration.

At the same time, it is necessary to note that the changes in optical transmissivity and diffraction efficiency that occur are not smooth with increasing As content. It seems to be reasonable to connect such behaviour with some discontinuity of atomic arrangement with increasing As content. Our recent study of composition dependence of Raman bands in a-$As_x Se_{1-x}$ supports this suggestion.

3.5 Summary

The molecular structure of a-Se differs from that of the majority of inorganic glasses on the scale of medium-range order. Apparently, Se chains form a structure similar to the structure of linear polymers on a scale of $\sim 1 \, nm$. The dependence of the transmissivity and diffraction efficiency on the irradiation energy density shows two qualitatively different regions. Below the energy density threshold E_{th}, only small changes in the local structure of the system can be detected. In the low-energy region, the transmissivity varies reversibly (transient changes) with exposure. The observed behaviour may be explained qualitatively by associating this with alternation of deep defect states. Above E_{th}, the changes were attributed to crystallisation transformation. In addition, we have detected the successive phases in such a transition which is a threshold phenomenon.

It has been shown that Raman scattering spectra of a-$As_x Se_{1-x}$ alloys change non-smoothly with composition in the region of bond-stretching modes. Certain extrema in various physical properties exist at composition range 6–12 at% As. The presence of this topological threshold is established by direct evidence, such as peculiarities in the compositional dependence of the Raman vibration modes of glasses. These peculiarities are caused by the transition from a chain-ring-like structure to chain-like structure with increasing degree of crosslinking.

By correlating the changes observed in optical properties under increasing irradiation energy density with insights into molecular structure gained from Raman scattering studies, we have shown how the structure transforms with time in a-$As_x Se_{1-x}$ recording media.

Laser-induced optical changes at room temperature involve two phenomena essentially different in their origin: transient reversible changes (PD) and irreversible changes (photocrystallisation) with gross structural reorganisation. For high values of energy density, the Raman spectrum has pronounced crystallisation-related changes. Our explanations are based on the assumption that the radiation pumps the material from an amorphous state towards the crystalline state through the formation of small clusters, which coalesce to form large clusters, attaining microcrystallite size at high energy density levels.

References

1. A. Madan, M.P. Shaw, *The Physics and Applications of Amorphous Semiconductors* (Academic, Boston, MA 1988)
2. A. Feltz, *Amorphous Inorganic Materials and Glasses* (VCH, Weinheim, Germany 1993)
3. K. Tanaka, In *Encyclopedia of Materials* (Elsevier, Oxford 2001) p.1123
4. S.O. Kasap, In *Handbook of Imaging Materials*, 2nd ed., ed. by A.S. Diamond, D.S. Weiss (Marcel Dekker, New York 2002), p.329, and references therein
5. S.O. Kasap, J.A. Rowlands, J. Mater. Sci. Mater. Electron. **11**, 179 (2000)
6. Z. Borisova, *Glassy Semiconductors* (Plenum, New York 1981)
7. N.F. Mott, E.A. Davis, *Electronic Processes in Non-Crystalline Naterials*, 2nd edn. (Oxford University Press, Oxford 1979)
8. J. Schottmiller, M. Tabak, G. Lucovsky, A. Ward, J. Non-Cryst. Solids **4**, 80 (1970)
9. A.E. Owen, W.E. Spear, Phys. Chem. Glasses **17**, 174 (1976)
10. V.I. Mikla, D.G. Semak, A.V. Mateleshko, A.R. Levkulich, Sov. Phys. Semicond. **21**, 266 (1987)
11. V.I. Mikla, D.G. Semak, A.V. Mateleshko, A.R. Levkulich, Sov. Phys. Semicond. **23**, 80 (1989)
12. K. Tanaka, Rev. Solid State Sci. **2/3**, 644 (1990)
13. J. Dresner, G.B. Stringfellow, J. Phys. Chem. Solids **29**, 303 (1968)
14. J.P. De Neufville, In *Optical Properties of Solids – New Developments*, ed. by D.O. Seraphin (North-Holland, Amsterdam, 1975), p.437

15. H. Fritzsche, In *Insulating and Semiconducting Glasses*, ed. by P. Boolchand, Chap.10 (World Scientific, Singapore, 2000)
16. M. Cardona, *Light Scattering in Solids*, Springer Tracts in Modern Physics (Springer, Berlin, 1975)
17. R.J. Nemanich, G.A. Connell, T.M. Hayes, R.A. Street, Phys. Rev. B **18**, 6900 (1978)
18. V.I. Mikla, Ph.D. Thesis (Odessa State University, Odessa, 1984)
19. A.A. Baganich, V.I. Mikla, D.G. Semak, A.P. Sokolov, Phys. Status Solidi B **166**, 297 (1991)
20. M. Gorman, S.A. Solin, Solid State Commun. **18**, 1401 (1976)
21. M.H. Brodsky, M. Cardona, J. Non-Cryst. Solids **31**, 81 (1978)
22. P.J. Carroll, J.S. Lannin, Solid State Commun. **40**, 81 (1981)
23. P.J. Carroll, J.S. Lannin, J. Non-Cryst. Solids **35/36**, 1277 (1980)
24. J. Jakle, In *Amorphous Solids: Low-Temperature Properties*, ed. by W.A. Phillips (Springer, Berlin, 1981) p. 135
25. V.I. Mikla, Ph.D. Thesis, Institute of Solid State Physics (Academy of Sciences, Kiev, 1984)
26. V.K. Malinovsky, V.N. Novikov, A.P. Sokolov, Fiz. Khim. Stekla **15**, 331 (1989)
27. V.K. Malinovsky, A.P. Sokolov, Solid State Commun. **57**, 757 (1986)
28. G. Lucovsky, J. Non-Cryst. Solids **97/98**, 155 (1987)
29. H. Richter, Z.P. Wang, L. Ley, Solid State Commun. **39**, 625 (1981)
30. V.M. Lyubin, *Non-Silver Photographic Processes* (Khimiyia, Leningrad 1984)
31. K. Tanaka, N. Odajima, Solid State Commun. **43**, 961 (1982)
32. R.T. Phillips, J. Non-Cryst. Solids **70**, 359 (1985)
33. V.A. Bagrynskii, V.K. Malinovsky, V.N. Novikov, L.M. Puschaeva, A.P. Sokolov, Fiz. Tverd. Tela **30**, 2360 (1988)
34. T. Mori, S. Onari, T. Arai, J. Appl. Phys. **19**, 1027 (1980)
35. S. Onari, K. Matsuishi, T. Arai, J. Non-Cryst. Solids **74**, 57 (1985)
36. J.C. Phillips, J. Non-Cryst. Solids **43**, 37 (1981)
37. M.F. Thorpe, J. Non-Cryst. Solids **57**, 355 (1983)
38. K. Tanaka, Phys. Rev. B **39**, 1270 (1989)
39. T. Wagner, S.O. Kasap, Philos. Mag. B **74**, 667 (1996)
40. V.I. Mikla, J. Phys. Condens. Matter **9**, 9209 (1997)
41. V.L. Averianov, A.V. Kolobov, B.T. Kolomiets, V.M. Lyubin, Phys. Status Solidi. A **57**, 81 (1980)
42. K. Tanaka, A. Odajima, Solid State Commun. **43**, 961 (1982)
43. V.I. Mikla, D.G. Semak, A.V. Mateleshko, A.A. Baganich, Phys. Status Solidi. A **117**, 241 (1990)
44. P. Boolchand, M. Jin, D.I. Novita, S. Chakravarty, J. Raman Spectrosc. **38**, 660 (2007)
45. E. Ahn, G.A. Williams, P.C. Taylor, Phys. Rev. B **74**, 174206 (2006)
46. R. Zallen, M.L. Slade, A.T. Ward, Phys. Rev. B **3**, 4257 (1971)
47. M. Abkowitz, R.C. Enck, Phys. Rev. B **25**, 2567 (1982)
48. I. Abdulhalim, R. Besserman, Solid State Commun. **64**, 951 (1987)
49. H.M. Yang, W.Z. Wang, S.K. Min, J. Non-Cryst. Solids **80**, 503 (1986)

Chapter 4
Trap Level Spectroscopy in Amorphous Selenium-Based Semiconductors

An informative and relatively simple spectroscopic technique, namely measurement of thermally stimulated depolarisation currents, is considered in this chapter for the study of defect states in the mobility gap of amorphous Se-based semiconductors.

Thermally stimulated conductivity (TSC) and thermally stimulated depolarisation currents (TSDCs) are well-known techniques for obtaining data on the trapping levels of crystalline semiconductors, and the techniques have also been successfully applied to amorphous semiconductors [6–18]. TSDC provides researchers today with an active arena of technological as well as fundamental study. On the fundamental front, TSDC provides a powerful framework for understanding the bandgap structure and properties of amorphous materials. The main attraction of TSDC as an experimental method for the study of defects in high-resistance solids was, for many years, their apparent simplicity. Trapping levels in the bandgap determine the fundamental electronic properties of both phases. In conventional TSC measurements difficulties arise if the thermal excitation of equilibrium carriers becomes comparable with the excitation of trapped nonequilibrium carriers. In this situation the TSC signal appears in the best case only as a shoulder on the dark current–temperature curve. One of the main difficulties in observing TSC in amorphous semiconductors is the small magnitude of the TSC currents [11, 14–18]. In most chalcogenide materials – and this may be considered a universal property – the TSC measurements yield no peak. As for the amorphous ones, it is important to note that presently no universal method is known to detect the entire spectrum of trapping levels in the mobility gap. This is why investigators employ several complementary methods. Among these, those that are convenient for the study of shallow and deep trapping levels, respectively, should be distinguished. For the former, nonisothermal relaxation techniques and time-of-flight measurements seem to be "suitable," whereas for the latter xerographic spectroscopy is most frequently employed. Each method has its advantages and disadvantages and often it may be useful to apply several of the methods listed above to the same specimen. TSDCs allow a relaxation processes attributed to relatively shallow trapping levels to be studied. The disadvantage in measuring TSDCs is the fact that the signals detected are very small and, sometimes, can be observed only in a relatively narrow temperature interval. At the same time, the advantage is that the TSDC method is inherently more sensitive than other methods and the resolution is usually much better. In addition, these measurements may be classified as nondestructive.

V.I. Mikla and V.V. Mikla, *Metastable States in Amorphous Chalcogenide Semiconductors*, Springer Series in Materials Science 128,
DOI 10.1007/978-3-642-02745-1_4, © Springer-Verlag Berlin Heidelberg 2010

In the present section we review the results of the extensive studies of defect states in the mobility gap of amorphous As- and Sb-containing chalcogenide semiconductors by relaxation techniques. To extract typical features, elemental selenium and simple compositions with relatively low contents of arsenic and antimony are taken as examples where possible. We will try to attribute TSDC peaks to charge-carrier release from the respective trapping levels in the bandgap of these materials.

Usually, two basic types of relaxation techniques are used:

- Isothermal relaxation: the perturbation is implemented at a constant temperature.
- Nonisothermal relaxation: the system is perturbed at a sufficiently low temperature to reduce the probability of a new statistical equilibrium being established. Subsequently, the temperature is increased according to a well-controlled heating program $T(t)$ and the relaxation of the system can be monitored as a function of temperature and time.

In the following the emphasis is on nonisothermal relaxation.

The occurrence of TSDC during a thermal scan of a previously excited (perturbed) material is probably the most direct evidence we have for the existence of electronic trap levels in the bandgap of these materials. The main attraction of TSDC and related techniques as experimental methods for studying the trapping levels in high-resistance semiconductors is their apparent simplicity. A TSDC spectrum (for historical reasons frequently referred to as a "glow curve") usually consists of a number of more or less resolved peaks a plot of current versus temperature, which, in most cases, may be attributed to a species of traps.

Since the probability of escape of carriers from trapping states is proportional to T, the location of a glow peak on the temperature scale provides information on the value of the thermal activation energy E_t. Hence, a glow curve represents a spectrum of energy required for carriers to be released from various traps in the material.

4.1 Thermally Stimulated Depolarisation Currents in Amorphous Chalcogenides: Background

At the very beginning, as far as amorphous materials are concerned, it must be remembered that the validity of a pure trapping model is still a moot point. In such nonperiodic structures the immobilisation of charge carriers for long periods does not necessarily indicate the presence of traps; a simple alternative explanation is that the carriers are slowly moving by an inherently low-mobility process, such as hopping. At the same time, it should be noted that although chalcogenide glasses may have traps distributed throughout the mobility gap, it appears justifiable to use the single-trap approach to calculate the trapping parameters (especially activation energy) of the materials under study. In addition, such a model based on a simplified version of the relaxation kinetics will be the commonly employed procedure in other complicated cases.

Consider first a single trap level with energy in the mobility gap of a p-type semiconductor. We assume that initially the sample is at a uniform temperature, low enough to prevent thermal emission of holes from their traps. Increasing the temperature according to a heating program leads to release of trapped carriers until thermodynamic equilibrium is reached again at some higher temperature. Further, we assume the density of equilibrium carriers is less than that of those released from traps and, neglecting diffusion and carrier recombination or generation, the excess charges can be considered as uniformly distributed and concentrated in narrow layers close to the electrodes (barrier-type polarisation), if at least one of the semiconductor–electrode interfaces can be considered as fully blocking (insulated electrode). The discharge of a photoelectret, that is, an amorphous material previously polarised by the photoelectret effect, may be expressed in the form

$$Q(T) = Q_0 \exp\left[\frac{4\pi e \mu}{\chi v_{\mathrm{T}}} \int_{T_0}^{T} p(T) \mathrm{d}T \right], \tag{4.1}$$

where Q_0 is the initial charge, μ the microscopic mobility, χ the permittivity of the material, p the concentration of free holes in the valence band and v_{T} the heating rate. The current induced in the measuring circuit by the internal field of the photoelectret during heating (irreversible scan) is described by

$$I(T) = \frac{4\pi e \mu}{\chi} Q_0 p(T) \exp\left[-\frac{4\pi e \mu}{\chi v_{\mathrm{T}}} \int_{T_0}^{T} p(T) \mathrm{d}T \right]. \tag{4.2}$$

We may write
$$p(T) = p_0(T) + p_{\mathrm{T}}(t) \tag{4.3}$$
with
$$p_0(T) = C \exp\left[-\frac{\Delta E}{2kT} \right], \tag{4.4}$$

where p_0 is the pre-exponential factor of the dark conductivity and k is the Boltzmann constant.

For states distributed in energy or for N discrete states very close to each other, (4.3) transforms to

$$P(T) = C \exp\left[\frac{\Delta E}{kT} \right] + \sum_{i=1}^{N} p_{ti}(T). \tag{4.5}$$

The ratio of p_0 to p_T in (4.3) determines which of two factors, namely equilibrium or nonequilibrium (due to emission from traps) carriers, dominate in the relaxation process. That is, the depolarisation current has two maxima: one is related to release of carriers from traps, while the origin of the other lies in the change of conductivity with temperature [14–18]. Although only one of the peaks contains information about trap parameters, it is possible to discriminate between simultaneously occurring processes, for example thermally stimulated depolarisation and thermally stimulated dielectric relaxation.

For trap depth to be defined, it is necessary to solve the rate equations [19]

$$\frac{d\,(m+p)}{dt} = \frac{p\,(t)}{\tau} \tag{4.6}$$

and

$$\frac{dm\,(t)}{dt} = -\alpha\,(t)\,m\,(t) + \gamma\,[M - m\,(t)]\,p\,(t), \tag{4.7}$$

where

$$\alpha = \gamma N_v \exp\left[-E_t/kT\right]. \tag{4.8}$$

The expressions (1–5) are similar to those describing TSC processes obeying first-order kinetics and represent an asymmetrical glow curve, the amplitude of which is a function of heating rate.

Detailed solutions have not been discussed in the literature so far. The time dependence of the rate equations (4.4) and (4.5) is replaced by the temperature dependence via the heating program, which is taken to be linear $T\,(t) = T_0 + v_T t$, where v_T is the heating rate. Re-trapping can be neglected in a sample at high electric fields. From (4.4) and (4.5) we have

$$\frac{dp}{dT} = \frac{dm}{dT} + \frac{p}{v_T \tau} \tag{4.9}$$

$$\frac{dm}{dT} = \frac{\alpha\,(T)\,m\,(T)}{v_T} \tag{4.10}$$

$$p_t\,(T) = \tau m\,(T)\,\gamma N_v \exp\left[-E_t/kT\right], \tag{4.11}$$

where $\gamma = \langle V \rangle S$, V is the average thermal velocity of charge carriers, S the capture cross section (for neutral, attractive or repulsive centres), and N_v the density of states in the valence band. It is obvious that the only parameters markedly affecting the peak amplitude and position will be the heating rate in combination with the characteristic frequency factor and the activation energy.

Equation (4.2) is solved numerically. It was found to be a good approximation for a wide range of physically reasonable trapping and other parameters, typical of a wide range of amorphous semiconductors:

$$1.0\,\text{eV} \leq \Delta E \leq 2.0\,\text{eV},$$
$$10^{23}\,\text{cm}^{-3} \leq C \leq 10^{30}\,\text{cm}^{-3},$$
$$\text{ca. } 10^{-5}\,\text{cm}^2\,\text{V}^{-1}\,\text{s}^{-1} \leq \mu \leq \text{ca. } 10^{-3}\,\text{cm}^2\,\text{V}^{-1}\,\text{s}^{-1},$$
$$0.2\,\text{eV} \leq E_t \leq 0.8\,\text{eV},$$
$$10^7\,\text{s}^{-1} \leq \gamma N_v \leq 10^{13}\,\text{s}^{-1}.$$

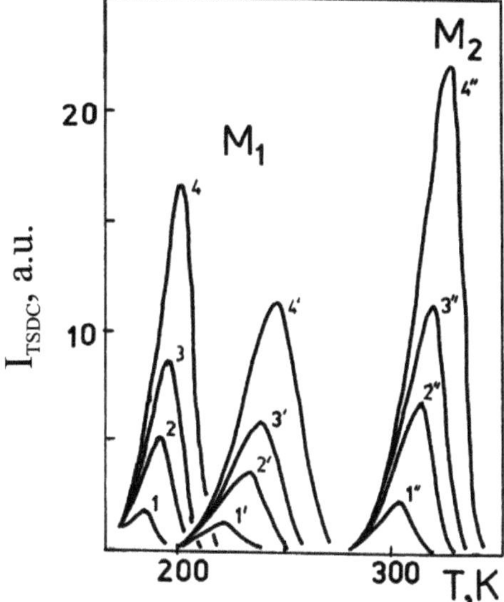

Fig. 4.1 Numerical solutions of (4.2) for the case $E_t = 0.4\,\text{eV}$ and $\gamma N_v = 10^7$ and $10^9\,\text{s}^{-1}$ (curves 1–4 and $1'$–$4'$, respectively). Heating rates are 0.1, 0.3, 0.5 and $1.0\,\text{K}\,\text{s}^{-1}$ (curves 1–4, respectively). Also shown are the dielectric relaxation currents (maximum M_2) for the same heating rates (curves $1''$–$4''$, respectively)

The capture cross section of a trap is largely determined by its charge state. Values reported in the literature ([3.1] and references therein) span the range 10^{-15}–$10^{-12}\,\text{cm}^2$ for Coulomb-attractive centres, 10^{-17}–$10^{-15}\,\text{cm}^2$ for neutral centres and down to $10^{-22}\,\text{cm}^2$ for Coulomb-repulsive centres.

Some examples of numerical solutions of (4.2) are shown in Fig. 4.1. It should be emphasised here that numerical solutions of (4.2) exhibit a greater variety of shapes, peak positions and magnitudes than shown.

Although two peaks of comparable amplitude are presented (Fig. 4.1), only the first, denoted M_1, is actually related to the carriers' release from the trap; the second, denoted M_2, is connected with dark conductivity variation with temperature (dc conductivity-determined).

Rate equations predict the influence of the heating rate on the amplitude as well as on the position of the peak: when the heating rate increases, the initial polarisation has to be released in a shorter time. As a result, the peak increases and shifts to a higher temperature. In fact, both M_1 and M_2 change with heating rate in a similar manner, but only for M_1 does the temperature location depend on the values of m and E_t at a given heating rate (Fig. 4.2).

Such behaviour of the TSDC peak is the basic condition for the activation energy of a trap to be determined correctly. On the other hand, the dc conductivity-determined relaxation peak shifts to lower temperatures with the bandgap and/or pre-exponential factor decreasing (Fig. 4.3).

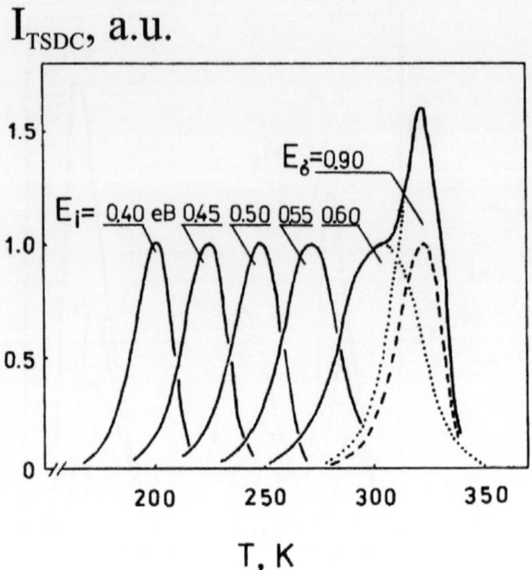

Fig. 4.2 TSDC maximum dependence on energy location of trapping level E_i ($\gamma N_v = 10^9\,\mathrm{s}^{-1}$, $C = 10^{25}\,\mathrm{cm}^{-3}$)

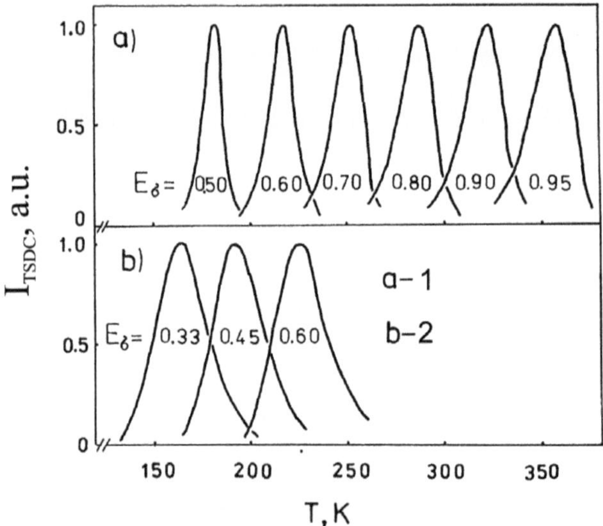

Fig. 4.3 Dependence of dielectric relaxation maximum on activation energy E_σ [eV] of the dc conductivity. (**a**) and (**b**) denote various activation energy values of the dc conductivity

Thus, the variation of the conductivity value, mobility gap and activation energy of the trap let us establish the range where TSDC peaks directly related to traps can be observed "for sure." For example, one can detect traps with $E_t \leq 0.6\,\mathrm{eV}$ only if the dielectric relaxation peak is located close to room temperature.

Table 4.1 Values of E_σ and E_{TSDC}^a for $Cu_x (As_2Se_3)_{1-x}$ glasses

Composition	E_σ [eV]	E_{TSDC} [eV]
As_2Se_3	0.91	0.89
$Cu_{0.05} (As_2Se_3)_{0.95}$	0.60	0.60
$Cu_{0.10} (As_2Se_3)_{0.90}$	0.45	0.40
$Cu_{0.15} (As_2Se_3)_{0.85}$	0.33	0.30

[a] E_σ and E_{TSDC} were determined from the temperature dependence of the dc conductivity and TSDC curves, respectively

Very illustrative of this statement are TSDC experimental results on $Cu_x (As_2Se_3)_{1-x}$ glasses. Significant increase of dark conductivity and decrease of activation energy with Cu content is characteristic of these glasses. Accordingly, the conductivity-determined depolarisation maximum shifts to lower temperatures (Table 4.1). As for the peak associated with traps, this can be observed only in $Cu_{0.05} (As_2Se_3)_{0.95}$. In $Cu_x (As_2Se_3)_{1-x}$ glasses with $x > 0.05$ the TSDC maximum observed is simply the result of two peaks overlapping – that determined by the equilibrium (dark) conductivity and that associated with trapping levels.

The presence of a maximum in the TSDC curve that is caused by dc conductivity was also confirmed in the extensive article by Agarwal [14], in particular for chalcogenide glasses. Moreover, it was shown that the TSDC curves may contain more than one peak, the position and amplitude of which depends on the size and relative geometrical arrangement of the electrodes with respect one to another.

The model presented by us has been found to fit the experimental data satisfactorily in many cases, particularly for high-resistance amorphous chalcogenide semiconductors. We have compared calculated and experimentally observed depolarisation currents for $As_{0.5}Se_{0.5}$. Two relatively broad maxima of comparable amplitude, curve shape, temperature location and similar behaviour with variation of the heating rate were present in the TSDC plot. The former (peak temperature T_{M1}) is attributed to trapping levels while the latter (peak temperature T_{M2}, such that $T_{M2} > T_{M1}$) is caused by dielectric relaxation current. In fact, one of the major problems of thermally stimulated current measurements is to unequivocally determine the physical origin of the observed current peaks. This is not an easy task and a great deal of controversy still surrounds the interpretation of TSDC experimental data for most of the materials tested. In our case the origin of the second peak on the depolarisation curve is strongly supported by the appearance of single peak, close to room temperature, in the thermally stimulated capacitor discharge experiment. This involves filling the states located near the Fermi level at some temperature (e.g., room temperature) by applying a strong field and subsequent cooling to a lower temperature with the field applied. Then the field is removed and the sample heated in the usual manner. The current, measured during heating, is the dielectric relaxation current. This contains information only on electronic states giving rise to dark (equilibrium) conductivity.

4.2 Thermally Stimulated Depolarisation Currents in Se-Based Amorphous Semiconductors: Experimental Details

4.2.1 Sample Preparation

The sample preparation is always the same, independent of the concrete measuring technique applied. Usually, the particular geometry of the sample may be adapted to the circumstances desired. Especial attention was paid to homogeneous field distribution in the sample. This allows an unambiguous interpretation of the experimental results. To avoid any leakage currents, the sample surfaces were carefully cleaned. Blocking electrodes were formed simply by pressing small metal lamellae (or foil) onto well-polished bulk samples (amorphous film). Highly insulating spacers were put under the lamellae. Thin lead wires were attached to the electrodes, which allow a good electrical contact. The experiments are carried out with different electrode materials and structures with insulated electrodes. In the latter case, insulating layers are inserted between the sample and the measuring electrodes. TSDC measurements were carried out on bulk glass (sample thickness 0.5 mm) and amorphous films (thickness 1–10 μm) samples. The heating rate was varied by application of different voltages across the heater. Typical heating rates were in the range 0.1–$1.0\,\mathrm{K\,s^{-1}}$. At lower heating rates the current signal becomes very small, while at higher heating rates the temperature gradient inside the bulk sample causes signal distortion.

4.2.2 Experimental Arrangement

As a general rule, the detection efficiency expected in conventional TSC experiments will be no more than 0–15% for carrier drift. It is possible, however, to significantly increase the efficiency of TSDC by using an insulating electrode adjacent to one side of the sample (insulating foil). Since the insulating electrode blocks any charge exchange, all image charges previously induced at the noncontacting electrode (due to carriers de-trapping from gap states) will be released during the TSDC run; it will be possible also to observe the current resulting from dc conduction. We initially assumed a well-defined sandwich-cell configuration consisting of a sample that is insulated from the metallic electrodes. The experimental apparatus used for recording TSDC is classical; it is essentially composed of a cryostat with sample holder, voltage supply, heater and a current detector. Standard metal–glass cryostats are used in low temperature studies of TSDC. The temperature of the films is controlled by mounting a heater inside the sample holder, and measured with a thoroughly calibrated copper–constantan thermocouple. Electrical leads and feed-throughs are designed for minimal leakage currents and stray capacitance. The sample was provided with a digitally controlled voltage supply of extreme stability and low noise level.

TSDC experiments are customarily analyzed assuming the sample behaves "ohmically," that is, the contacts do not introduce an inhomogeneous distribution of the electric field or carrier density and a uniform bulk density of carriers extends throughout the entire sample. Experiments were carried out in such a way to as minimise injection effects. The contact configuration was typical for TSDC experiments. Because the currents through the sample are in almost all cases extremely small, we used a sensitive dc ammeter (model U1-15, detection limit $\leq 10^{-15}$ A) with linear output signal. The simplest way to obtain a record of TSDC is an $X-Y$ recorder, which displays current and temperature. The equipment for the extraction of trap-spectroscopic information may be connected with devices for electronic data processing. The experimental errors in determination E_t are less than 2%.

In the TSDC experiment considered here, a sample is cooled to a low temperature (100 K) and illuminated with 3×10^3 lx light for a time t_p (4 min) in the presence of an applied DC field $\left(E = 5 \times 10^4 \, \text{V cm}^{-1}\right)$. Then the light and voltage are switched off, the structure is "short-circuited" and after a delay period, necessary for sample relaxation (to reach equilibrium between the free and the trapped carriers), the sample is heated in the darkness at a constant rate while the TSDC is measured. We preferred TSDC experiments because of the absence of noise due to a voltage source, and the strongly reduced influence of the intrinsic conductivity.

4.3 Thermally Stimulated Depolarisation Currents in Se-Based Amorphous Semiconductors: Results

4.3.1 TSDC in Pure Selenium

Conventional TSC in glassy selenium exhibits monotonic temperature dependence without distinct "structure," which is characteristic for the crystalline analogues. The characteristic peaks attributed to traps are usually absent in TSC versus temperature plots. Such behaviour is typical for chalcogenide glassy semiconductors. Although TSC exceeds the dark conductivity, the absence of a well-defined structure of the TSC does not allow the corresponding trapping levels to be identified [17–19]. Therefore, other methods of thermal activation spectroscopy are needed, for example measurement of TSDC.

The TSDC in glassy selenium samples starts from the "photoelectret" state. The latter is most strongly expressed at blocking electrodes. Thus, and in order to separate the TSDC peak from the dc conduction, we placed a highly insulating layer between the glass specimen and metal electrodes in the measuring cell. Two different materials were used as the dielectric insulator: cleaved mica sheet and Teflon (polytetrafluoroethylene). The two dielectric materials gave essentially the same results.

TSDC measurements on glassy selenium samples display a well-shaped, nearly symmetrical, peak at 150 K. It is a general fact in chalcogenide glasses (amorphous

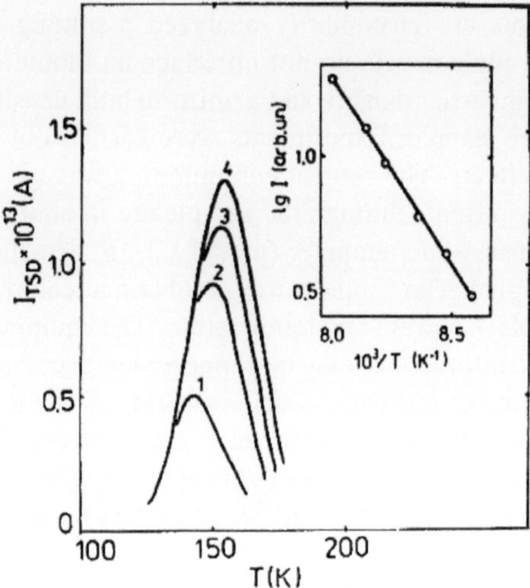

Fig. 4.4 Variation in the TSDC for various heating rates in amorphous selenium. Heating rates 0.08′0.17, 0.23 and 0.44 K s^{-1} (curves 1–4, respectively). The inset shows the Arrhenius plot of TSDC for curve 1

materials) that the peaks extend over a wide temperature range compared to their crystalline analogues. In addition, the peaks are flatter and more symmetrical than expected from the simple expressions (Sect. 4.1). At lower heating rates, the peak shifts to lower temperatures and is reduced in height, as expected (Fig. 4.4).

The principal trap parameter, the activation energy, can be easily calculated from a single TSDC experiment by means of some characteristic elements of the peak, such as its half-width, inflection point or initial current rise. The most useful one, and in fact the most frequently exploited, is undoubtedly the initial rise method (Garlick and Gibson) [20], which is always easily applicable to a previously cleaned peak.

The initial rise method of Garlick and Gibson (for details see [9]) is based on the fact that, because the integral term in the $J_D(T)$ function is negligible at $T \ll T_M$, where T_M is the temperature of the maximum, the first exponential dominates the temperature rise of the initial current, so that

$$\ln J_D \approx \text{const.} - \frac{E}{kT} \qquad \text{(Arrhenius shift)} . \qquad (4.12)$$

The activation energy can be determined by plotting $\ln J_D$ against $1/T$. In this approximation, a straight line is obtained, the slope of which gives $-E/k$.

Indeed, this method has several advantages, as it necessitates neither a linear heating rate nor a precise knowledge of the absolute temperature (it is readily seen, for example, that a thermal gradient of 5 K in a sample leads to a relative error less than 3% at 300 K). It is, however, somewhat limited because use of only the initial part of the curve is permitted, which may force one into the region where the uncertainty in background current is important. On the other hand, the method is still applicable for overlapping relaxation, even if the relaxation modes cannot be isolated by cleaning, provided that a series of partial heating runs, each separated by rapid cooling, are performed up to gradually increasing temperatures that span the whole temperature range of the spectrum.

The activation energy associated with the peak at 150 K has been calculated by the above method, that is, by plotting $\ln J_{\mathrm{D}}$ against $1/T$. Such a plot is shown in the inset (Fig. 4.4). The activation energy obtained by this method is 0.22 eV.

An independent check of the correctness of the activation energy can be obtained by measuring the TSDC at different heating rates [21]. The so-called varying heating rate methods are based on the shift of the TSDC maximum with heating rate [22]. Various ways of plotting the results have been proposed. From the equation for the temperature of the peak maximum

$$T_{\mathrm{M}} = \left[\frac{E}{k} v_T \tau_0 \exp\left(\frac{E}{kT_{\mathrm{M}}} \right) \right]^{1/2} \tag{4.13}$$

it is readily seen that the activation energy for an Arrhenius shift can be determined by using two heating rates v_{T1} and v_{T2} and measuring the corresponding values T_{M1} and T_{M2} of the maximum

$$E = \frac{kT_{\mathrm{M1}}T_{\mathrm{M2}}}{T_{\mathrm{M1}} - T_{\mathrm{M2}}} \ln\left(\frac{v_{T1}T_{\mathrm{M2}}^2}{v_{T2}T_{\mathrm{M1}}^2} \right). \tag{4.14}$$

A better procedure is that proposed by Hoogenstraaten. A number of heating rates are used and T_{M} is determined as a function of v_{T}. A plot of $\ln\left(T_{\mathrm{M}}^2/v_{\mathrm{T}}\right)$ against $1/T_{\mathrm{M}}$ should yield a straight line, from the slope of which the activation energy can be calculated. The success of these methods for the analysis of TSDC curves will thus depend on how much the position of a given peak can be shifted. They will, therefore, be applicable only when the activation energies are small.

Finally, simplified expressions for the current density can be used to calculate a family of theoretical TSDC curves – nomograms – which are then compared to the experimental data.

When no information is available on the distribution in E_{t}, and this is nearly always the case, only the step-heating technique and, to a lesser degree, the varying heating rate methods can be expected to be used with some success.

It is necessary to note that the measurement of the carrier density by TSDC experiments is not a simple matter and inhomogeneities in both the carrier and the electric field distributions along the sample may seriously affect the validity of the results obtained. In most cases these undesirable effects can be minimised.

Note also that the systematic variation of specimen thickness (between 50 μm and 500 μm) and using different materials as the dielectric insulator yielded nearly the same peak position and activation energy. This suggests strongly that disturbing contact effects were absent. Moreover, no difference was observed between specimens open-circuited or short-circuited during cooling, nor did the time interval between irradiation and subsequent heating appear to have any affect on the observations. Therefore, it appears that the TSDCs considered here are the true TSDCs associated with thermal release of nonequilibrium carriers from localised gap states. The trap of activation energy 0.22–0.24 eV acts as a shallow trap.

In addition to the low-temperature peak a peak at ca. 270 K (not shown in Fig. 4.4) is also present in the TSDC curve. The ratio of amplitudes of these two peaks is about 5. It is important to note here that a dielectric relaxation current peak (in the depolarisation of thermoelectret state experiments) at ca. 270 K is also observed. The activation energy determined for this peak is ca. 0.60 eV. Importantly, this parameter depends on electrode configuration and sample thickness. Therefore, a peak at ca. 270 K may be attributed simply to the change in dark conductivity with temperature.

4.3.2   TSDC in As $(Sb)_x$ Se$_{1-x}$ alloys

Arsenic addition to amorphous selenium, even in relatively low amounts (ca. 1 at%) as for glassy compounds, causes a change in the shape of the TSDC curve. The transition from pure amorphous Se to $As_{0.01}Se_{0.99}$ is accompanied by a broadening of the corresponding TSDC curve and the appearance of an additional maximum (Fig. 4.5).

The main maximum on the depolarisation curve is shifted with respect to those observed in a-Se and is located in the range 190–200 K. This maximum is the result of overlapping of two maxima with $T_M \approx 191$ K and $T_M \approx 203$ K. In addition, a shoulder is observed at 160 K. The shoulder and the peak position observed in pure selenium correlate. For the separation of the overlapping maxima, the method of partial thermal cleaning was used. This method is based on release of carriers from traps by sequences heating to temperatures higher then T_M of the corresponding maximum, rapid cooling and heating again. The result obtained after applying such a procedure to $As_{0.01}Se_{0.99}$ is shown in Fig. 4.6. Values of activation energy

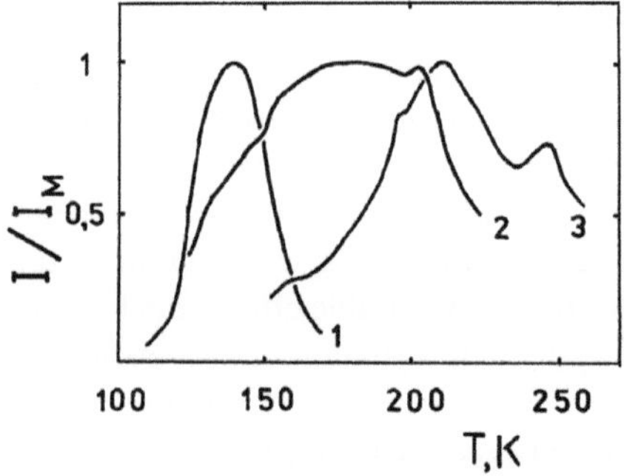

Fig. 4.5   TSDCs in glassy Se (1), $As_{0.01}Se_{0.99}$ (2) and $As_{0.15}Se_{0.85}$ (3)

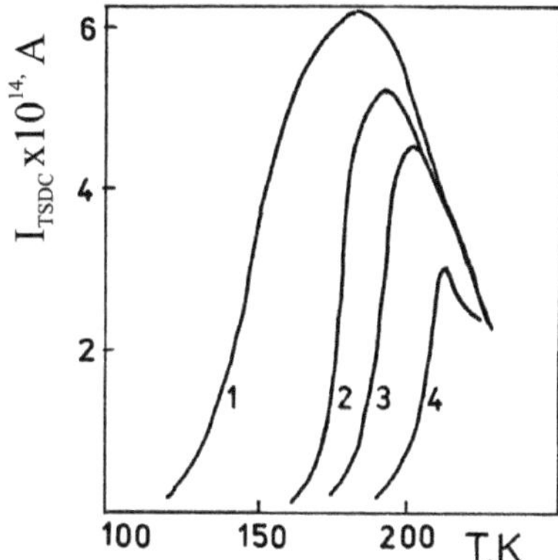

Fig. 4.6 Application of the thermal cleaning method of TSDC peaks to $As_{0.01}Se_{0.99}$. (1) Initial TSDC curve. The temperature in successive heating cycles is 166 K (2), 180 K (3) and 199 K (4)

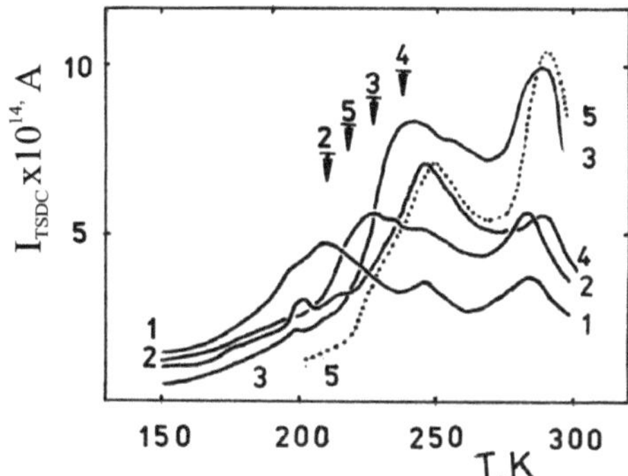

Fig. 4.7 TSDC curves of $As_{0.15}Se_{0.85}$ samples (1). Curves 2–5: Polarised samples were heated under applied dc voltage to the temperatures shown by the arrows

calculated by the initial rise method are in the range 0.25–0.45 eV. Continuous distribution of localised states in this range was confirmed by increasing activation energy in the sequence in curves 2–4 (Fig. 4.6.). Further increase of As addition (up to 6 at%) does not change the shape of the TSDC curve: the latter remains similar to that observed in $As_{0.01}Se_{0.99}$.

In Fig. 4.7, the TSDCs observed in $As_x Se_{1-x}$ for As content 8–15 at% is shown. A peak at $T_M \approx 210$ K dominates in the depolarisation curve. In addition, a peak at 244 K and a shoulder at 197 K are discernable. Release of carriers from shallow states and subsequent trapping on deeper ones transforms the TSDC into curves 3 and 4 (Fig. 4.7). Activation energy determined for peaks with $T_M \approx 210$ K and 244 K is $E_{t2} = 0.35$ eV and $E_{t3} = 0.45$ eV. These states are energetically distributed: $\Delta E_{t2} = 0.05$ eV and $\Delta E_{t3} = 0.1$ eV.

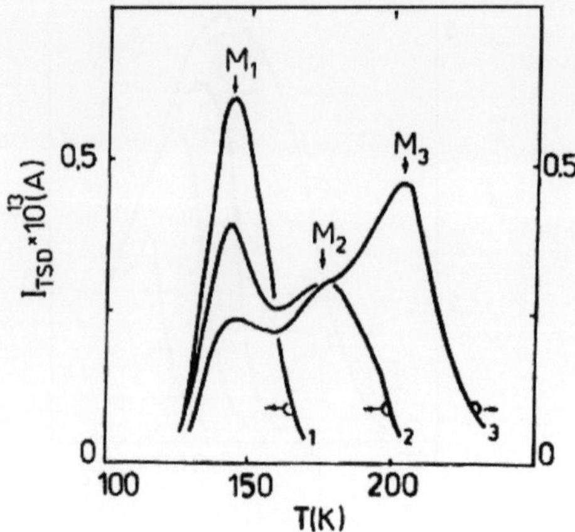

Fig. 4.8 TSDC spectra of Sb_xSe_{1-x}, $x = 0$ (1), 0.01 (2), 0.05 (3)

For TSDC in Sb_xSe_{1-x} alloys, we have three peaks M_1, M_2 and M_3, whose temperatures T_{M1}, T_{M2} and T_{M3} (145, 175 and 195 K, respectively) are close enough to each other that the peaks overlap (Fig. 4.8). The TSDC method allows quick and easy resolution of the overall spectrum, in our case peaks 1, 2 and 3. There are two ways in which this can be done.

The first method, which can be used with all nonisothermal techniques, was developed by Greswell and Perlman [22]. Suppose, for example, we have two peaks 1 and 2, whose temperatures are close enough to each other as to overlap. After the whole TSDC curve has been obtained, a second thermal cycle is started. However, instead of recording the discharge at one time, we first discharge the lower temperature peak 1 by heating to $T \ll T_{M2}$, then we cool the sample again and finally obtain the discharge of peak 2, which is "pure" or nearly "pure."

The second cleaning method, specific to TSDC measurements and which may be applied in our case, is due to Bucci et al. [23]. It consists of first polarising the material at a temperature T_a $(T_M < T_a)$ for a sufficient interval of time: the resulting TSDC curve will then show only peak 1. This is repeated at temperature T_b $(T_M < T_a < T_b)$, allowing peak 2 to be isolated in a similar way. Such an effective way of separating the above-mentioned peaks is illustrated in Fig. 4.9. The activation energies determined by these methods were $E_{t1} = 0.22$ eV, $E_{t2} = 0.34$ eV, and $E_{t3} = 0.45$ eV. Note that some of the above traps in the materials studied were "visible" in time-of-flight and xerographic experiments [5, 17, 19].

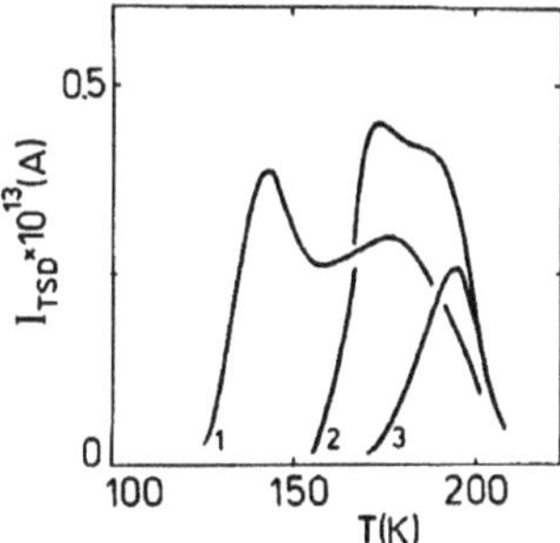

Fig. 4.9 Thermal cleaning procedure (for TSDC) applied to $Sb_{0.01}Se_{0.99}$. Curve 1: The whole TSDC curve. For curves 2 and 3, $T_a = 165\,K$ and $T_b = 195\,K$ (see the text)

4.4 Summary

The results taken as a whole reveal the existence of at least three different trap species in the bandgap of $Sb(As)_x Se_{1-x}$ noncrystalline semiconductors. These species are located at 0.22, 0.34, and 0.45 eV below the conduction band edge and control the electron transport properties of the material. It seems that Sb and As introduce a new set of detectable charge carrier traps. Space considerations preclude a detailed discussion of similar trapping levels in other high-resistance amorphous chalcogenides.

Finally, we are convinced that the future development of TSC and TSDC methods as tools to investigate the electronic properties of amorphous semiconductors will benefit from similar careful studies by means of time-of-flight and xerographic measurements.

References

1. M. Popescu, J. Non-Cryst. Solids **352**, 887 (2006)
2. V. Kolobov (ed.), *Photoinduced Metastability in Amorphous Semiconductors* (Wiley-VCH, Weinheim, Germany, 2003)
3. J. Singh, K. Shimakawa, *Advances in Amorphous Semiconductors* (Taylor and Francis, London, 2003)
4. G. Lucovsky, M. Popescu, *Non-Crystalline Materials for Optoelectronics* (INOE, Bucharest, 2004)
5. S.O. Kasap, *Handbook of Imaging Materials*, 2nd ed., ed. by A.S. Diamond, D.S. Weiss (Marcel Dekker, New York, 2002)
6. D. Kumar, S. Kumar, Chalcogenide Lett. **1**, 49 (2004)
7. E. Skordeva, J. Optoelectron Adv. Mater. **3**, 437 (2001)
8. V.F. Zolotaryov, D.G. Semak, D.V. Chepur, Phys. Status Solidi **21**, 437 (1967)
9. P. Braunlich, P. Kelly, J.P. Fillard, *Thermally Stimulated Relaxation in Solids*, ed. by P.Braunlich, Top. Appl. Phys. **37** (Springer, Berlin, 1979)
10. Yu. Gorokhovatsky, G. Bordovskij, *Thermally Activated Current Spectroscopy of High-Resistance Semiconductors and Dielectrics* (Nauka, Moscow, 1991) (in Russion)

11. R.A. Street, A.D. Yoffe, Thin Solid Films **11**, 161 (1972)
12. B.T. Kolomiets, V.M. Lyubin, V.L. Averjanov, Mater. Res. Bull. **5**, 655 (1970)
13. T. Botila, H.K. Henish, Phys. Status Solidi A **36**, 331 (1976)
14. S.C. Agarwal, Phys. Rev. B **10**, 4340 (1974)
15. S.C. Agarwal, H. Fritzsche, Phys. Rev. B **10**, 4351 (1974)
16. P. Muller, Phys. Status Solidi A **67**, 11 (1981)
17. V.I. Mikla, Thesis, Institute of Physics, National Academy of Sciences, Kiev (1998)
18. A.A. Kikineshi, V.I. Mikla, I.P. Mikhalko, Sov. Phys. – Semicond. **11**, 1010 (1977)
19. V.I. Mikla, I.P. Mikhalko, Yu.Yu. Nagy, J. Phys. Condens. Matter **6**, 8269 (1994)
20. G.F. Garlick, A.F. Gibson, Proc. Phys. Soc. London **60**, 574 (1948)
21. A.H. Bohun, Can. J. Chem. **32**, 214 (1954)
22. R.A. Greswell, M.M. Perlman, J. Appl. Phys. **41**, 2365 (1970)
23. C. Bucci, R. Fieschi, G. Guidy, Phys. Rev. **148**, 816 (1966)

Chapter 5
Photoinduced Effects on Electronic Metastable States

The effect of pre-excitation with light of bandgap energy on trapping and thermal generation have been examined in selenium and selenium-rich As–Se alloy films by several techniques. Results suggest that excess-carrier trapping and dark-carrier generation are controlled by deep defect centres whose population can temporarily be altered by photoexcitation.

5.1 Steady-State Photocurrents

There is a strong evidence to suggest that most of the reported photoinduced optical and structural changes affect the bandgap states. Several research groups have devoted considerable efforts to experiments that probe the gap states near mobility edges and within the mobility gap (see, e.g. [1] and references therein). This is not a simple matter; no single experiment gives complete and unequivocal information and obviously it is necessary to bring together data from a wide range of measurements, stationary as well as transient.

First of all we consider the associated changes in the photoelectronic properties of the samples. The spectral characteristics of photoconductivity of the samples display a red shift after irradiation. Such behaviour of the photoconductivity is not surprising, as it is in full agreement with the shift in the absorption edge. Additionally, the photoconductivity decreases after photodarkening (PD) [2]. The decrease may be attributed to the creation of new defect states or alteration of the existing localised states.

Illuminance–current (lux–ampere) characteristics of annealed films are characterised by a sublinear dependence $I_{ph} \propto E^n$, where I_{ph} is photocurrent, in a wide light intensity range, n being approximately equal to 0.6 for As_2Se_3. After darkening, the index n increases.

The temperature dependence of the steady-state photocurrent is generally quite similar to those observed widely in chalcogenides [3]. Figure 5.1 shows that photoconductivity is an activated process and varies exponentially with $1/T$. The activation energy from the plot of $\log I_{ph}$ versus $10^3/T$ turns out to be 0.31 eV. Also plotted in the figure is the variation in dark conductivity with temperature, which

V.I. Mikla and V.V. Mikla, *Metastable States in Amorphous Chalcogenide
Semiconductors*, Springer Series in Materials Science 128,
DOI 10.1007/978-3-642-02745-1_5, © Springer-Verlag Berlin Heidelberg 2010

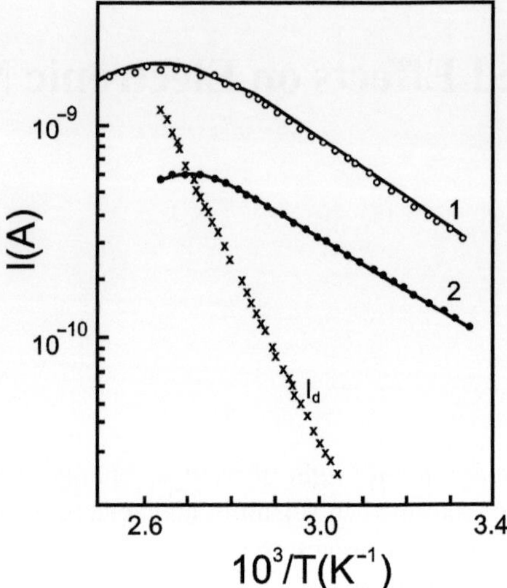

Fig. 5.1 Steady-state photocurrent in a-As$_2$Se$_3$ sample at 700 nm illumination with approximate intensity of 10^{14} photons cm^{-2} s^{-1} as a function of inverse temperature (*curve 1*). Also shown is the effect induced by laser irradiation ($\lambda = 633$ nm) at $T = 100$ K (*curve 2*)

is also an activated process. Since it is generally accepted [3] that in chalcogenide semiconductors recombination will be mediated by the charge defect states, we may analyze the temperature dependence of the steady-state photocurrent in terms of an energy scheme involving acceptor-like levels in the gap. The low-temperature "bimolecular" slope ΔE_b, we believe, represents half the distance between the valence band and acceptor-like level. The observed differences in ΔE_b with exposure ($\Delta E_b \approx 0.22$ eV in irradiated samples) therefore indicate that illumination influences the gap-state density in the lower half of the bandgap.

The above changes in the photocurrent of amorphous chalcogenides may be related to specific changes in electronic gap states, which act as trapping and recombination centres and which therefore limit the photoconductivity.

5.2 Light-Induced Effects on Photocurrent Transients

The most unique and intriguing features of chalcogenide vitreous semiconductors are the photoinduced changes appearing as a nearly parallel shift of the optical absorption edge to lower energy (PD effect) after irradiation with bandgap light. Such irradiation also causes changes in a wide variety of other properties — mechanical, physicochemical, photoelectronic, etc. [4]. The latter provides a valuable opportunity for the evaluation of light-induced changes of the gap states. However, during the last few years there has been very limited systematic study of photoinduced effects on states, especially of deep-lying states in the bandgap. Most

PD studies have focused upon binary and ternary chalcogenide alloys and relatively little is known about the characteristics of elemental and chalcogen-rich glasses. On the other hand, elemental and chalcogen-rich amorphous semiconductors serve as useful model systems for studying the influence of PD on physical properties. We consider in the following photoinduced changes of deep levels in selenium layers with arsenic additions up to 20%.

Currently two classes of transient photocurrent experiments can be distinguished. On the one hand, there are the unipolar methods. The most common is the time-of-flight (TOF) experiment developed by Spear [5]. In the TOF method the carriers are created near one electrode of sandwich configuration. This configuration makes it possible to distinguish electron and hole motion. The carrier mobility is determined by the position of the break in the current pulse. Therefore, precise knowledge of the concentration of the transiting carriers is not necessary. To observe the transit pulse in the TOF experiment, the electrode must form a blocking contact, that is, it must be able to inject carriers. To prevent distortion of the field, the excess charge introduced should be less than CV, where C is the capacitance of the sample and V is the voltage across the sample.

On the other hand, there is the bipolar measurement of transient photocurrent (due to electrons and holes) performed in the gap-cell configuration (see, e.g. [6,7]). In these experiments the sample is illuminated uniformly in the direction transverse to the electrodes, creating an excess conductance G. Injecting contacts supply the excess current $I_{ph} = GV$ for a bias voltage V. There is no build-up of charge at the electrodes. As a result, signal averaging with repetitive pulses is easy. When analyzing coplanar transient photocurrents, the different drift mobility of electrons and holes, together with recombination, must be taken into account.

First of all we consider results of TOF measurements on Se-rich amorphous chalcogenides [8]. The sample films used in all TOF studies were typically 5–50 μ m thick. These were prepared by vacuum evaporation of $As_x Se_{1-x}$ alloys at a fixed rate of ca. $1 \mu m \, min^{-1}$ onto glass substrates. The substrates were coated with SnO_2 or Al blocking contacts. All samples were annealed at room temperature for ca. 100 h and stored in the dark for at least ca. 48 h before measurement. Where sandwich geometry was required (photocurrent transients), semitransparent gold contacts were slowly evaporated onto the free surface of the specimens. Bandgap illumination was provided by a mercury lamp and a He-Ne laser. The transient transport measurements were performed using a conventional (small signal) TOF technique. The drift of the sheet of carriers initially photoexcited at the Au contact by a nitrogen laser pulse (5 ns duration, 337.1 nm wavelength) is time-resolved in a uniform applied field. All measurements were performed in the current mode, in which the time constant of the external circuit iss kept short compared to the transit time to ensure constant voltage across the sample during the carrier drift. When dispersion entirely masks the transit time cusp we define a statistical transit time from a double-logarithmic plot of the algebraically decaying current.

Two types of xerographic measurements will be discussed, namely dark decay and cycled-up xerographic residual voltage [1,9]. The former experiments are carried out by charging the surface of the sample from a corotron device and measuring

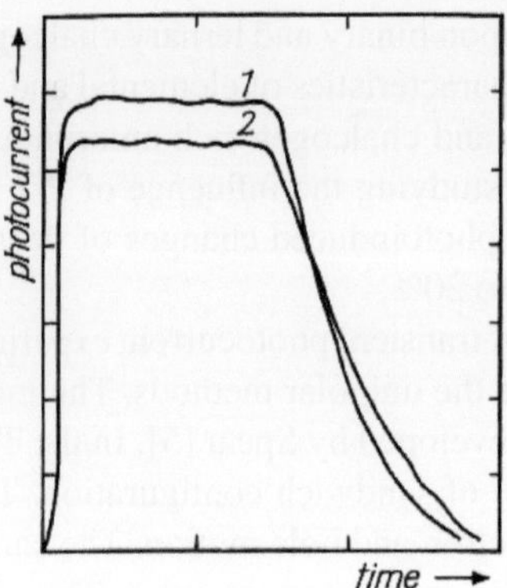

Fig. 5.2 A typical oscilloscope trace of the TOF photocurrent electron signal in (1) dark-rested and (2) exposed a-Se ($d = 4.2\,\mu\text{m}$, $E = 14\,\text{V}\,\mu\text{m}^{-1}$, $T = 293\,\text{K}$, $0.5\,\mu\text{s}\,\text{div}^{-1}$)

the surface potential under an electrostatic probe. In the latter experiments (cycled-up residual voltage) the surface voltage following photodischarge in the xerographic cycle is monitored as a function of the number of cycles.

At room temperature, the amorphous Se (a-Se) films exhibit very slight but definite PD. The effect consists of a decrease of transmission ($\approx 2\%$ in magnitude) after illumination, which was spontaneously restored when excitation was switched off. It was found that the restoration period increases with As concentration. In other words, the photoinduced changes are transient (dynamic) and not permanent.

In order to observe the effect of pre-illumination (exposure conditions were chosen identical to those that cause transmission changes) on microscopic parameters of states, measurements of the transport characteristics were performed. We consider first the carrier drift and space charge as it develops in $\text{As}_x\text{Se}_{100-x}$ alloy films with $0 \leq x \leq 15$ at%, in which transport of holes and electrons is observable [3, 8]. Figure 5.2 shows a typical nondispersive trace of the transient electron current in dark-rested and pre-exposed pure selenium. The room temperature value of the drift mobility, $\mu^e \approx 6 \times 10^{-3}\,\text{cm}^2\,\text{V}^{-1}\,\text{s}^{-1}$, is in remarkably good agreement with reported values and seems to be unaffected by irradiation. We observe only a small decrease of the current and a slightly changed shape of the signal. There is a close correlation between the recovery of optical parameters and the current in exposed samples. For example, the transmission of irradiated a-Se approaches its virgin value after ca. 7 min. The same time is necessary for relaxation of the electron drift transient. Note that in the case of the hole photocurrent transients about 2 min is sufficient to restore the original signal.

Arsenic alloying results in a decrease of both the electron and hole mobilities μ^e and μ^h and substantially changes the shape of the hole transients. At low As concentrations (0–2 at%) the hole transit pulses are well defined and retain the shape of pure Se. In the range $2\,\text{at}\% \leq x \leq 6\,\text{at}\%$ As the hole response is absent and only a lifetime-limited signal is observed, in accordance with the reported [8, 10] behaviour. On raising the As concentration above 6 at%, the hole signal reappears,

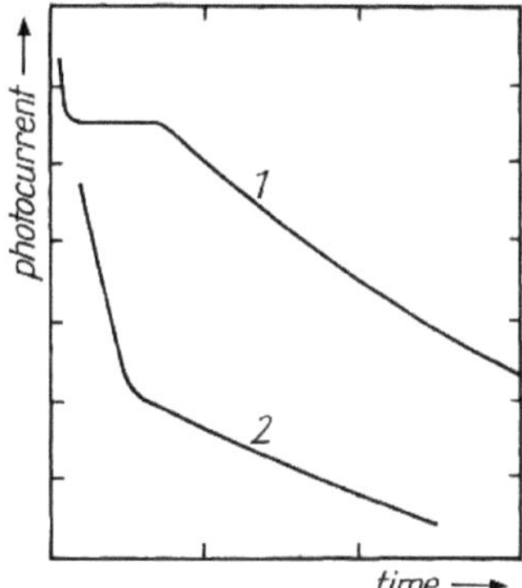

Fig. 5.3 Oscilloscope trace of the TOF photocurrent hole signal in (1) dark-rested and (2) exposed a-$As_{0.10}Se_{0.90}$ ($d = 4.2\,\mu m$, $E = 14\,V\,\mu m^{-1}$, $T = 293\,K$, $0.5\,\mu s\,div^{-1}$)

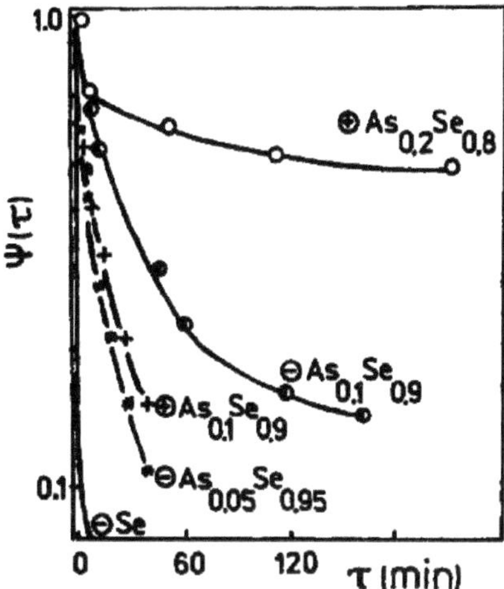

Fig. 5.4 The relaxation function for hole (circled plus signs) and electron (circled minus signs) metastable traps

but now it consists of an initial fast response, a quasi-stationary region (plateau), and a long featureless tail typical of current transients observed in As_2Se_3. Bandgap irradiation of the sample leads essentially to a decrease of the current level, whereas the transit time t_T remains unchanged. If after light exposure the sample is allowed to rest in the dark, signal recovery takes place [2, 8] (Fig. 5.3).

Figure 5.4 shows the relaxation function $\psi\,(\tau)$ of irradiated As_xSe_{100-x} alloy films, which is determined as follows. The sample irradiated with light of photon energy $h\nu \geq E_{opt}$ for a fixed time was allowed to rest in the dark for a time τ (experimental variable). The current signal $I = f\,(t)$ was recorded as a function of

$\psi(\tau)$, where $\psi(\tau)$ was determined as

$$\psi(\tau) = \left[I_{\mathrm{r}} - I_{\tau}\right] / \left[I_{\mathrm{r}} - I_{\mathrm{p}}\right] \tag{5.1}$$

where I_{r} and I_{p} are the current values of dark-rested and exposed samples, respectively, immediately after illumination (current traces were chosen at $t = t_T$). Results for the relaxation function $\psi(\tau)$ were fully analogous to those obtained from charge collection experiments.

The increase of As content and/or lowering of the temperature delay the relaxation of both electron and hole transit pulses in photosensitised samples. At the same time, asymmetry of the relaxation functions for electrons and holes was clearly detected. From the data presented it is obvious that the transit time remains unchanged by PD for all investigated compositions. Consequently, shallow states, which control charge transport and define the activation energy E_μ ($E_\mu^{\mathrm{e}} \approx 0.33\,\mathrm{eV}$, $E_\mu^{h} \approx 0.25\,\mathrm{eV}$ in a-Se, $E_\mu^{h} \approx 0.4$ to $0.6\,\mathrm{eV}$ in As_2Se_3) of the mobility, should not undergo photoinduced changes. At the same time, the observed change of the current shape, the appearance of a long tail, and the decreasing signal height indicate photoinduced change of deep ($E_{\mathrm{t}} > E_\mu$) centres. Such behaviour is possibly related to enhanced carrier trapping by deep levels after photosensitisation. Figure 4.5 supports this suggestion. It is apparent from the figure that pre-illumination leads not only to a decreasing magnitude of the photocurrent transient (in this case for electrons in an $As_{10}Se_{90}$ alloy film), but also to an abrupt transformation of the signal character: from a pulse with well-defined transit time (curve 1, Fig. 5.5) to a featureless lifetime-limited pulse (curve 2, Fig. 5.5). The considerable decrease of transit pulse height was accompanied by a significant alteration of the

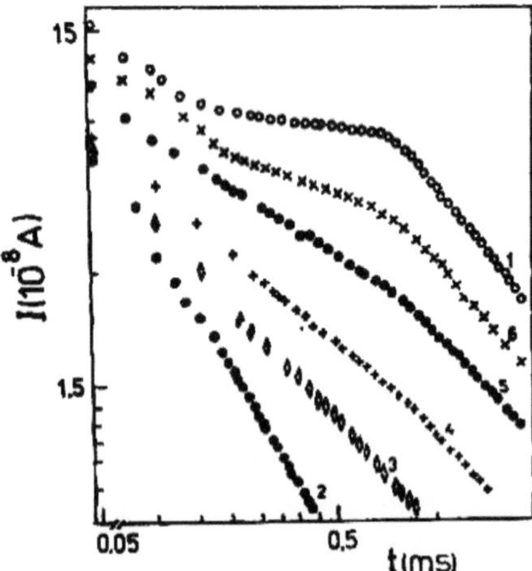

Fig. 5.5 Transient electron current in $As_{0.10}Se_{0.90}$ samples ($d = 4.4\,\mu\mathrm{m}$, $E = 3.7\,\mathrm{V}\,\mu\mathrm{m}^{-1}$, $T = 293\,\mathrm{K}$): (1) dark rested; (2) exposed; (3–6) exposed and then dark rested for 5, 10, 50, and 160 min, respectively

current-versus-time shape: the absolute value of the initial (pretransit) current slope $I \propto t^{-(1-\alpha)}$ in double-logarithmic representation (see the corresponding curves in Fig. 5.5). If, after pre-illumination, the film is allowed to relax in the dark, the previous (dark-rested) shape of the transient current is fully recovered, that is, pulse height and shape return to values characteristic of the unexposed specimen. It is thus found that charge-carrier lifetime with respect to deep trapping τ_L decreases by approximately one order of magnitude. For example, τ_L is less than 0.5 ms for electrons in the previously illuminated sample $As_{0.10}Se_{0.90}$.

In an earlier study of transport in annealed As_2Se_3 films we observed similar behaviour of photoinjected carrier transients under conditions of PD [4.2]. The main difference to the results presented here is that, in As_2Se_3 films, once changed by irradiation the optical and transport parameters never relax spontaneously to the equilibrium state. Restoration is possible only by annealing near the glass transition temperature. Further investigations show that bandgap light leads to a pronounced variation of the space charge developing in the sample. Figure 5.6 shows typical transient hole photocurrents induced in As_2Se_3 films before and after PD. When transit measurements are repeated, the current level diminishes progressively (current traces 1 and 2,3 and 4, Fig. 5.6). It is obvious that after PD the space charge build-up is significantly intensified.

Drift mobility experiments may be affected by two kinds of space charge effect. First, the presence of deep centres in the gap [5]. These centres gradually accumulate charge with each drift pulse, especially near the surfaces of a specimen, which leads to a decreasing internal field. As a result the transient current decreases in magnitude. Second, at high photoinjection levels ($Q > CV$, where C is the sample

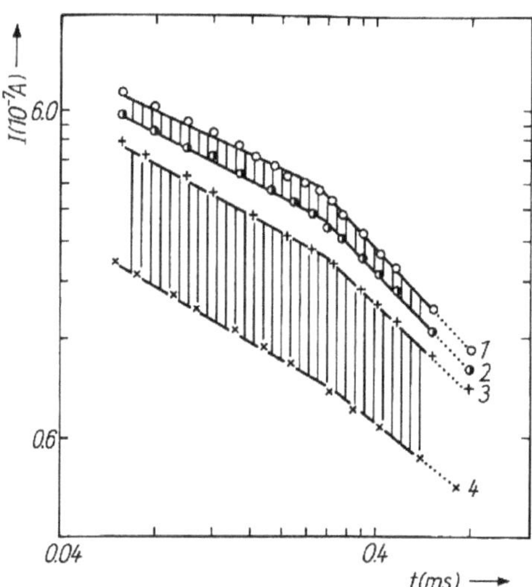

Fig. 5.6 Transient hole current in a $-\ As_2Se_3$. (1,2) Annealed sample; (3,4) photodarkened sample. The areas between (1) and (2) and (3) and (4) indicate space-charge accumulation during a ten-transit sequence with repetition frequency 10 Hz ($d = 3.4\,\mu m$, $E = 5.3\,V\,\mu m^{-1}$, $T = 293\,K$)

capacitance and V the applied field) space charge may be formed by the drifting packet itself. In our case the latter can be excluded because of the small value of injected charge ($Q \leq 0.1CV$).

Bulk space-charge creation is shown in Fig. 5.6 by the areas between curves 1 and 2 and 3 and 4. The estimated amount of charge accumulated by deep traps during a ten-transit sequence is about 1.3 to 2 times that at PD and approaches ca. $10^{14}\,\mathrm{cm}^{-3}$.

Consequently, shallow states, which control charge transport and define the activation energy E_μ^h (about 0.4 eV for As_2Se_3) of the mobility, should not undergo photoinduced changes. At the same time it should be stressed that the observed change in the current transient (namely its increased dispersion and decreased magnitude) indicate photoinduced changes of deep states with $E_i > E_\mu^h$. Such behaviour is probably related to enhanced carrier trapping by deep levels in photodarkened samples.

References

1. S.O. Kasap, *Handbook of Imaging Materials*, 2nd ed., ed. by A.S. Diamond, D.S. Weiss (Marcel Dekker, New York 2002)
2. V.I. Mikla, J. Phys., Condens. Matter **8**, 429 (1996)
3. N.F. Mott, E.A. Davis, *Electronic Processes in Non-Crystalline Naterials*, 2nd ed. (Oxford University Press, Oxford 1979)
4. K. Tanaka, *Encyclopedia of Materials* (Elsevier Science, Oxford 2001), p. 1123.
5. W.E. Spear, J. Non-Cryst. Solids **1**, 197 (1969)
6. J. Orenstein, M. Kastner, V. Vaninov, Philos. Mag. B **46**, 23 (1982)
7. V.I. Mikla, Ph.D. Thesis (Odessa State University, Odessa, 1984)
8. V.I. Mikla, D.G. Semak, A.V. Mateleshko, A.A. Baganich, Phys. Status Solidi A **117**, 241 (1990)
9. M. Abkowitz, R.C. Enck, Phys. Rev. B **25**, 2567 (1982)
10. A.E. Owen, W.E. Spear, Phys. Chem. Glasses **17**, 174 (1976)

Chapter 6
Deep Level Spectroscopy in Selenium-Rich Amorphous Semiconductors

Mobility measurements by the time-of-flight (TOF) method (Chap. 5) are particularly important but they must be augmented by others. Xerographic techniques, initially developed to characterise electrophotographic receptors, are widely applicable to the study of amorphous thin films and photoconductive insulator thin films. The high field (10^5–10^6 V cm^{-1}) caused by corona charging is applied to sample films, and the decay of the open-circuit surface potential is measured. The xerographic probe technique is a unique means of characterising electronic gap states. In particular, a map of states near mid-gap is determined by time-resolved analysis of the xerographic surface potential.

6.1 Xerographic Dark Decay and Photoinduced Effects

Analysis of the time- and temperature-dependent decay of the surface voltage on an amorphous film after charging but prior to exposure (xerographic dark decay) and of residual decay after exposure can (in combination) be used to map the density of states. The procedure is illustrated for a-$As_x Se_{1-x}$, where residual charge can be measured. Amorphous $As_x Se_{1-x}$ is found to be characterised by a relatively discrete gap state structure. These measurements readily discern photo- and thermostructural effects on gap state population. Thus during these structural transformations systematic variation in the number of localised states distributed throughout the mobility gap is observed. This observation is consistent with the view that native defects play a key role in photoelectronic behaviour of amorphous chalcogenides.

The illumination of amorphous $As_x Se_{1-x}$ films by light with energy near the optical gap also causes changes in basic electrophotographic parameters: dark discharge rate, initial charging potential, residual potential, and its dark decay rate. Typical dark discharge curves for film composition $As_{0.2} Se_{0.8}$ are shown in Fig. 6.1.

It is apparent that the initial charging voltage U_0 becomes smaller and the surface potential decay rate dU/dt increases after previous photoexcitation. It must be emphasised that, depending on composition, ordinary (changes of either U_0 or dU/dt) or complex (simultaneous changes of U_0 and dU/dt) photoinduced effects

V.I. Mikla and V.V. Mikla, *Metastable States in Amorphous Chalcogenide Semiconductors*, Springer Series in Materials Science 128, DOI 10.1007/978-3-642-02745-1_6, © Springer-Verlag Berlin Heidelberg 2010

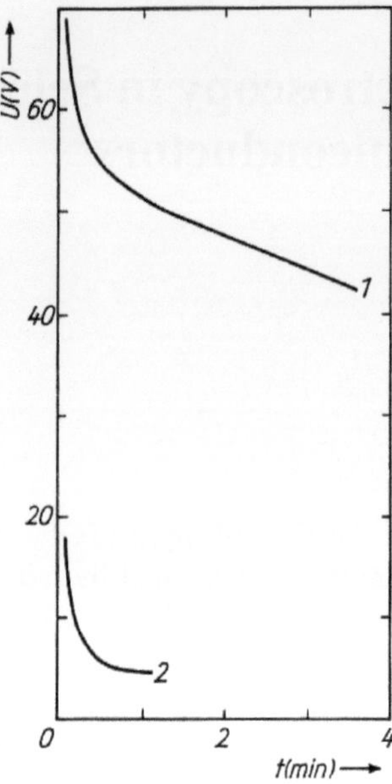

Fig. 6.1 Dark decay of surface potential in a-As$_{0.2}$Se$_{0.8}$: (1) Dark-rested sample; (2) after white light exposure 5,000 lx for 60 s

may be observable. The former take place in the range 2–6 at% As, whereas for the latter concentrations above 8 at% As are needed.

In pure selenium the photoinduced change of discharge rate is comparatively small ($\gamma \approx 1.06$). As the As concentration increases, γ initially becomes progressively larger, then for $x > 20$ at% it begins to fall, possibly because of the rising dark discharge rate in dark-rested samples. The parameters of dark discharge in dark-rested and photosensitised films are summarised in Table 6.1. Here $t_{U_0/2}$ is the decay half-life and $\Delta U/U_0 = \left(U_0 - U_0^*\right)/U_0$ the relative change of the initial charging potential. It is of interest to point out that the "memory" effect (the time interval during which the changes in pre-excited film parameters may be observable) is appreciably extended with increasing As content: from ca. 40 min in pure selenium to ca. 10 h in As$_{0.2}$Se$_{0.8}$.

Figure 6.2 shows the time evolution of the dark decay kinetics of the surface potential after photoexcitation. An increasing dark adaptation time, that is, the dark-resting time of an exposed film before it is charged to a certain potential and the potential decay recorded, causes a diminishing of the observable photoinduced changes. On the basis of the data in Fig. 6.2, by analogy to the photocurrent transients, the relaxation functions

$$\phi\left(\tau\right) = \frac{U_0 - U_0^{\tau}}{U_0 - U_0*} \tag{6.1}$$

Table 6.1 Surface potential dark decay parameters and their photoinduced changes in $As_x Se_{1-x}$ films[a]

As content [at%]	$dU/dt \left[\text{V s}^{-1}\right]$	$t_{U_0/2}$ [s]	γ[b]	$\Delta U/U_0$
0	6.7×10^{-2}	276	1.06	0.30
2	9.7×10^{-2}	150	1.20	–
5	1.4×10^{-1}	138	1.33	–
8	9.3×10^{-1}	19	1.61	–
10	1.22	17	2.3	–
15	1.12	18	3.5	0.50
20	1.4	15	1.9	0.51

[a] $U_0 = 37\,\text{V}$, $d = 10\,\mu\text{m}$, $T = 293\,\text{K}$, white light exposure 2,000 lx for 120 s.
[b] The ratio $\gamma = (dU^*/dt) / (dU/dt)$ (parameters of irradiated samples are marked by asterisks) is composition dependent

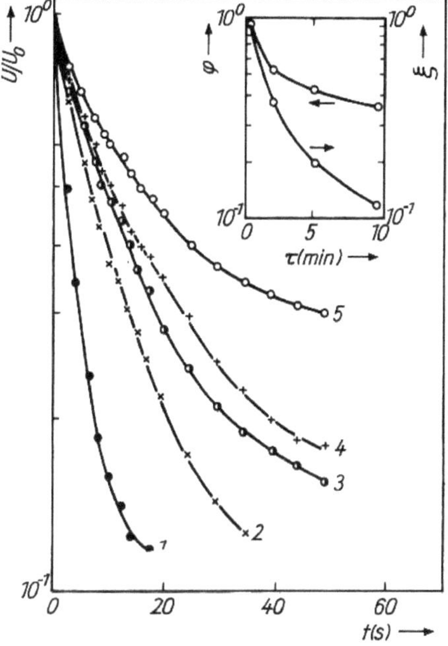

Fig. 6.2 Dark discharge rate varying with a-$As_{0.15}Se_{0.85}$ sample dark-resting time τ and the corresponding relaxation functions $\varphi(\tau)$ and $\zeta(\tau)$ (*inset*). $\tau = 0, 2, 5, 10,$ and 90 min for (1)–(5), respectively

and

$$\zeta(\tau) = \frac{(dU*/dt) - (dU^\tau/dt)}{(dU*/dt) - (dU/dt)}, \tag{6.2}$$

characterising the initial charging potential and the dark decay rate recovery, were estimated (see inset in Fig. 6.2).

It is of particular significance that "memory" effects are influenced not only by variation of composition, but also by electric fields. For example, Fig. 6.3 clearly

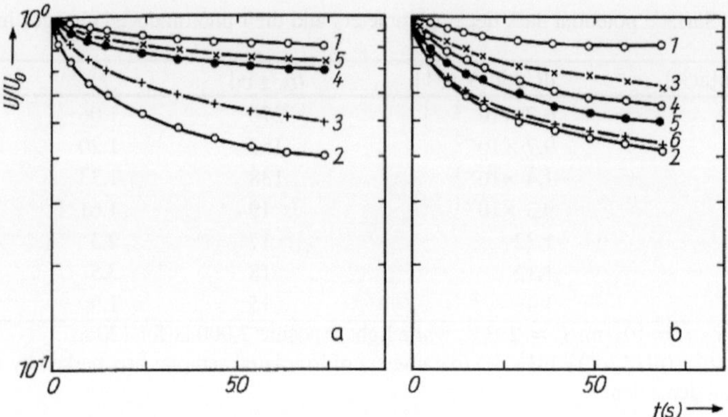

Fig. 6.3 Surface potential dark decay in a-$As_{0.1}Se_{0.9}$ films ($d = 10\,\mu m$). (**a**) (1) Dark-rested sample, (2) exposed sample, (3)–(5) exposed and then dark-rested for 5, 20, and 30 min, respectively. (**b**) (1) Dark-rested sample, (2) exposed sample, (3)–(6) exposed and dark-rested for 30 min in the presence of a surface potential of 115, 170, 215, and 520 V, respectively

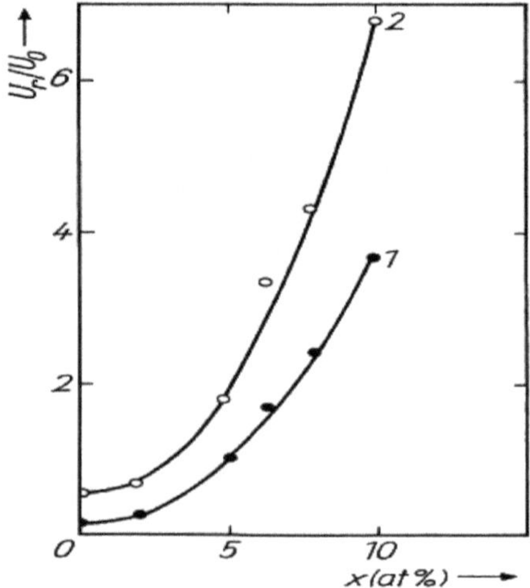

Fig. 6.4 Dependence of residual voltage on As content [at%] after (1) one and (2) five complete xerographic cycles

illustrates that the photoinduced change of the dark discharge rate may be successfully "frozen-in" by applying an electric field $E \approx 3 \times 10^5\,V\,cm^{-1}$ immediately after light exposure (compare curve 5 in Fig. 6.3a with curve 6 in Fig. 6.3b).

Dark decay curves of the surface potential in double-logarithmic representation display two distinct rate processes, $dU/dt \propto t^{-(1-K)}$ and $dU/dt \propto t^{-(1+K)}$, with cross-over from one regime to the other at $t = t_d$. Detailed analysis of the discharge process shows that t_d shifts to smaller values with increasing As content and after exposure (Fig. 6.4). Note that under low charging voltages U_0 the surface potential at time t_d was approximately half the initial value and that t_d increased with U_0. The features observed in the dark decay study of $As_x Se_{1-x}$ alloy films can be completely accounted for by a depletion discharge model [1–3]. So, the bulk process driving dark decay is emission and sweep-out of holes from states near mid-gap, leading to progressive formation of negative space charge. The rise in dark discharge rate and the shift of t_d with arsenic concentration and photoexcitation may be due to enhanced thermal generation of holes from deep centres. From the temperature dependence of t_d it is estimated that the emitting sites are located at 0.8–0.9 eV above the valence band mobility edge.

6.2 Residual Voltage in Se-Rich Photoreceptors

For the films under examination the residual voltage U_r increases with exposure. For example, previous photoexcitation leads to an increase in U_{r1} (first cycle residual voltage) for Se up to 3.7 V caused by a decrease in $\mu\tau_L$ to 1.3×10^{-7} cm^2 V^{-1}. Taking into account the invariance of electron mobility $\mu^e = 4.9 \times 10^{-3}$ cm^2 V^{-1} s^{-1} with light exposure, it is obvious that the lifetime reduction (from 5.4×10^{-5} to 2.6×10^{-5} s) is the only reason for photoinduced change.

During the continuous repetition of xerographic cycles, the residual voltage U_{rn} at the nth cycle was found to have the typical behaviour shown in Fig. 6.5.

The saturation residual potential U_{rs} provides [4, 5] a measure of the density of uniformly trapped carriers:

$$U_{rs} = Ned^2/2\varepsilon, \tag{6.3}$$

where ε is the dielectric constant. An increase in U_{rs} in pre-exposed films indicates photoenhanced accumulation of charge at deep centres. We obtain $N \approx 10^{15}$ cm^{-3}

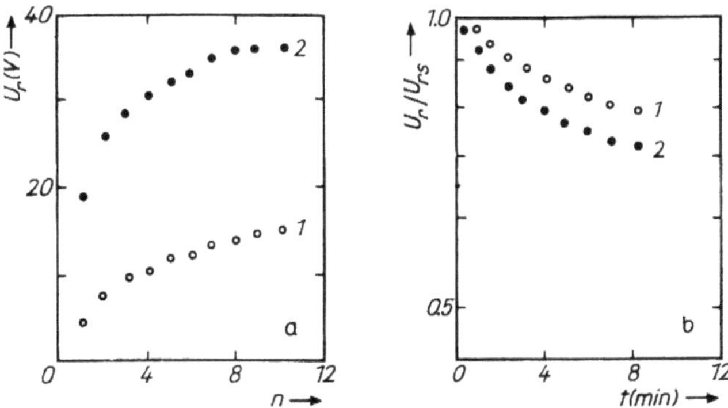

Fig. 6.5 The build-up in the residual voltage with (**a**) number of xerographic cycles and (**b**) isothermal room-temperature relaxation of saturated residual voltage in (1) dark-rested and (2) photoexcited a-$As_{0.02}Se_{0.98}$ films

and $N^* \approx 4 \times 10^{15}\,\mathrm{cm}^{-3}$ for dark-rested and pre-exposed $\mathrm{As}_{0.1}\mathrm{Se}_{0.9}$ films, respectively [6].

The residual potential decays to zero. This process is controlled by the spectrum of trap release times. The trap energies can be deduced from an analysis of isothermal residual potential decay curves using

$$U_{\mathrm{rs}} = \sum C_i \exp\left(-t/\tau_i\right), \tag{6.4}$$

where $1/\tau_i = \nu_i \exp\left(-E_i/kT\right)$ is the release time from the ith trap, ν_i is the frequency factor, and E_i is the trap depth. We find that deep levels in amorphous selenium reside at $E_i^{\mathrm{h}} = 0.85\,\mathrm{eV}$ and $E_i^{\mathrm{e}} = 1.0\,\mathrm{eV}$ for holes and electrons, respectively. Their depth becomes somewhat shallower with addition of As, for example, $E_i^{\mathrm{h}} = 0.80\,\mathrm{eV}$ and $E_i^{\mathrm{e}} = 0.90\,\mathrm{eV}$ in $\mathrm{As}_{0.1}\mathrm{Se}_{0.9}$. The more rapid decay of the residual voltage U_{rs} in pre-illuminated films relative to dark-rested film (Fig. 6.5b) indicates a slight decrease in the depth of those states. A comparison of room-temperature recovery in TOF and xerographic measurements demonstrates that relaxation of deep metastable centres in exposed films occurs on the same time scale. In other words, the deep gap centres that control the xerographic dark decay and residual voltage are, like the trapping centres discussed in TOF experiments, characteristically metastable.

Bandgap light can, in principle, have two distinct effects on the electronic structure of the mobility gap. Bandgap light can either introduce (generate) new localised states or initiate conversion of traps of small cross section to traps of larger cross section [5,6]. Consequently, the latter become accessible in deep-level spectroscopy only after irradiation. In that sense we may also consider such converted localised states as "new" localised states (created by irradiation).

References

1. A.R. Melnyk, J. Non-Cryst. Solids **35–36**, 837 (1980)
2. M. Abkowitz, J. Non-Cryst. Solids **97–98**, 1163 (1987)
3. M. Abkowitz, F. Jansen, A.R. Melnyk, Philos. Mag. B **51**, 405 (1985)
4. M. Abkowitz, R.C. Enck, Phys. Rev. B **25**, 2567 (1982)
5. S.O. Kasap, in *Handbook of Imaging Materials*, 2nd edn., ed. by A.S. Diamond, D.S. Weiss (Marcel Dekker, New York, 2002)
6. V.I. Mikla, J. Phys. Condens. Matter **8**, 429 (1996)

Chapter 7
Recombination Process and Non-Isothermal Relaxation of Low-Temperature Photoinduced Effects

Several complementary experimental techniques, including measurements of the transient response, steady-state photoconductivity and xerographic effect, have been employed to investigate photoinduced effects in amorphous arsenic selenide alloys. Results suggest that metastable deep defect centres can be altered by irradiation.

7.1 Recombination Process as it Appears in the Photocurrent Transients

Next, we turn our attention to the monomolecular recombination process, which appears in the photocurrent transients in Fig. 7.1.

This current–time plot (log I vs log t) compares the transients for non-photodarkened and photodarkened samples. The behaviour of the photocurrent is seen to be very similar to that of the current in the time-of-flight (TOF) experiment. In both cases the photocurrent makes a sharp transition from a power law with exponent less than one to a power law with exponent greater than one. Clearly, the monomolecular recombination time τ_{MR} plays the same role in the photocurrent experiment that the transit time t_T plays in the TOF experiment. In fact, Orenstein et al. [1] assumed that the photocurrent would decay this way.

As is clear from Fig. 7.1, τ_{MR} is slightly dependent on the light treatment. In principle, the carrier lifetimes are given by

$$\tau_{MR} \propto \frac{1}{v_0} \left(\frac{b_t N_L}{b_r N_{th}} \right)^{1/\alpha}, \tag{7.1}$$

where v_0 is the prefactor of the release rate from localised states (typically of the order of 10^{12}–$10^{13} s^{-1}$), b_t the trapping coefficient, N_L the factor of density of localised states, b_r the recombination coefficient, N_{th} the thermal equilibrium concentration (of recombination centres), and α the dispersion parameter characteristic of anomalous charge transport. There are two factors that both tend to make τ_{MR} greater as the sample is photodarkened. The first is the trap-limited mobility and the second is the concentration of recombination centres. From TOF data, the mobility

V.I. Mikla and V.V. Mikla, *Metastable States in Amorphous Chalcogenide Semiconductors*, Springer Series in Materials Science 128, DOI 10.1007/978-3-642-02745-1_7, © Springer-Verlag Berlin Heidelberg 2010

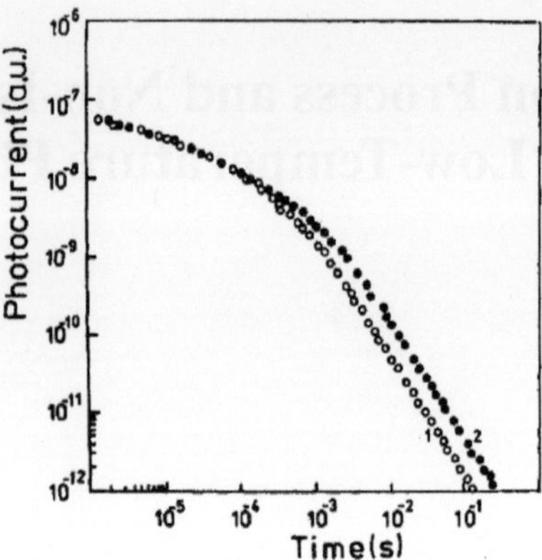

Fig. 7.1 The transient photocurrent from fully annealed (1) and photodarkened (2) a-As$_2$Se$_3$ gap cells after flash excitation ($\lambda_e = 690$ nm) (a.u., arbitrary units). The change in slope of the photocurrent transients is identified as τ_{MR}

remains unchanged with photodarkening. We speculate that the change in τ_{MR} can be explained by the second factor alone.

7.2 Non-Isothermal Relaxation of Low-Temperature Photoinduced Effects

A durable (permanent) photoinduced change known as photodarkening (PD) is retained unless the sample is annealed. In addition to this permanent change, transient (dynamical) photoinduced changes have also been observed in chalcogenide glasses. There are some indications that the dynamical changes are probably due to an alteration of metastable defects, which act as shallow traps for carriers generated by bandgap light. Reasonably, one can expect that the non-equilibrium carriers become immobilised for a long time in these states if the irradiation temperature is essentially lowered. It is assumed that the magnitude of PD is increased at low temperatures due to the additional (with respect to room temperature) component. The latter manifests itself as dynamical changes in transmissivity and refractive index at room temperature whereas it is transformed to a permanent change at $T = 100$ K. That is why it seemed plausible to study the non-isothermal relaxation of the low-temperature photoinduced effects.

Several transient and/or stationary methods are suitable for studying electronic and atomic relaxation. For this purpose, thermally stimulated currents (TSCs) and holographic grating techniques have been used. The latter can monitor even small changes in refractive index.

In the present investigation, a grating having a pitch $\Lambda = \lambda_e / [2 \sin (\theta/2)]$ is produced at $T = 100$ K by two interfering beams of wavelength λ_e (633 nm), intersecting at a sample surface with an angle θ ($\sim$40°). The diffraction efficiency

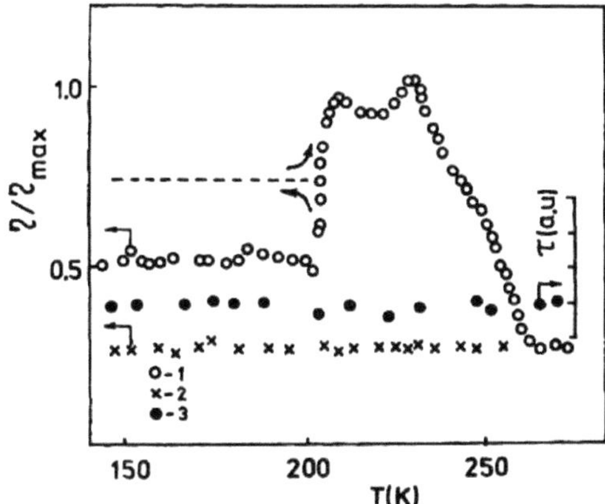

Fig. 7.2 The diffracted light intensity from gratings formed on an $As_{0.5}Se_{0.5}$ thin film at T $= 100$ K (*circles*). Also shown is the η vs T dependence after annealing at $T = 300$ K (*crosses*) (a.u., arbitrary units). Note that transmissivity of a fully annealed film is nearly temperature independent at $\lambda = 900$ nm (*filled circles*)

$\eta = I/I_0$, where I_0 and I are the corresponding intensities of the reading beam and of the first-order diffracted beam, is measured as a function of rising temperature. The reading beam of wavelength λ_r (900 nm) emitted by a 10 mW GaAs laser appears to be little affected by a change in absorption coefficient with rising temperature.

Figure 7.2 shows the typical response of light intensities diffracted from gratings formed on $As_{0.5}Se_{0.5}$ film. The η vs T data exhibit several stages: an initial invariance followed by a rapid rise at $T = 200$ K and then fall to values of $\eta/\eta_{max} = 0.3$ at $T = 300$ K. Finally, at $T \geq 420$ K (this high-temperature region is not shown in Fig. 7.2), η falls to zero, and the grating is completely erased. Note that the relative increase in η observed is about 50% with respect to its value at $T = 100$ K. This effect (namely "self-enhancement of holographic recording") may be of practical importance because it permits the diffraction efficiency to be essentially increased without additional treatment such as substrate metallisation and chemical etching. There are, however, factors limiting its applicability. Among them are the low-temperature holographic recording itself and inconveniences connected with such a procedure, which is a vanishingly small effect in a more suitable ($T \geq 300$ K) temperature range.

Now we turn our attention to data on TSCs. In order to be able to find the correspondence to data gratings, TSC measurements were performed on identical samples. The latter were illuminated at $T = 100$ K; the total exposure used was approximately the same as in the grating experiment. Figure 7.3 shows a well-shaped TSC peak with $T_{max} = 275$ K. The activation energy E_i associated with the peak has been calculated by the initial rise method. The E_i obtained by this method is about 0.4 eV.

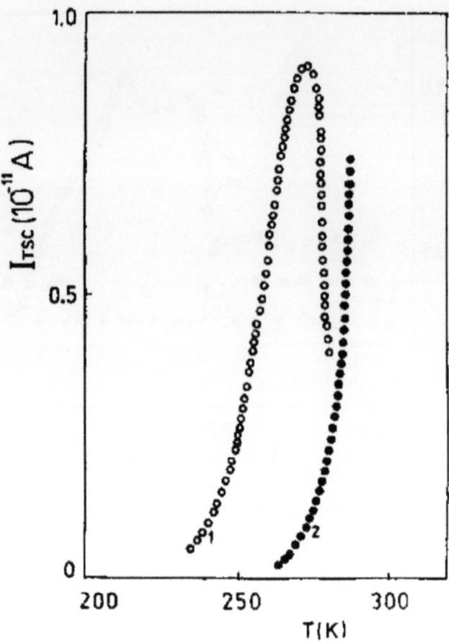

Fig. 7.3 Thermally stimulated current (TSC) curve in a-As$_{0.5}$Se$_{0.5}$ (1). The sample was cooled to 100 K and then exposed to laser illumination at 633 nm. After the whole TSC curve had been obtained, a second thermal cycle for an unexcited sample was started (2). As for Fig. 7.2, the heating rate is 0.1 Ks^{-1}

With regard to η versus T and I_{TSC} versus T dependence, note only that there is a certain correspondence; when the trapped carriers in the gap states of E_i ($E_i = 0.4$ eV) were emitted with a temperature rise the diffraction efficiency distinctly decreased. At the same time, the thermal decays of the two effects differ at higher temperatures. The TSC indicates almost complete annealing but a substantial fraction of η, induced at $T = 100$ K, is still present even at 300 K.

7.3 Possible Mechanism of Photoinduced Changes

The irreversible structural changes that occur in the conventionally prepared (slowly deposited) thin films upon optical illumination can be understood in terms of the same mechanism as described in [2–7]. These changes are associated with a photopolymerisation of As$_4$S$_4$ and S$_n$ molecular species. The resultant structure formed after such transformation is a three-dimensional network with nearly restored chemical ordering.

Rapid deposition, as indicated from the present Raman data, significantly enhances compositional as well as structural disorder in amorphous samples compared with samples with stoichiometric composition. An increase in As concentration leads to condensation of As$_4$S$_3$ and some other As-rich molecular species;

a disordered molecular solid is obtained. These As-rich compositions are unstable, and presumably phase separation occurs upon thermal annealing and light irradiation. Without going into further details of these irreversible effects in rapidly deposited films, it can be said that irreversible photoinduced structural changes in conventionally prepared amorphous films are quite different from a phenomenological point of view and also with respect to their origin.

We next discuss experimental investigations on reversible photoinduced effects in previously annealed films. Two sensitive experimental techniques, namely nuclear quadruple resonance spectroscopy [8] and Raman spectroscopy (see the Raman data presented in this volume), failed to detect any gross changes in the local bonding configuration. However, other structural probes, namely infrared absorption [9] and differential anomalous X-ray scattering [10], show that small changes do occur under illumination. A mechanism proposed by Elliott [11] to account for the light-induced changes in the structure describes PD in chalcogenide glasses in terms of intra-molecular and inter-molecular bond breaking and bond weakening. According to this model, the increase in homopolar bonding would have to be substantially larger (about 7% for As_2S_3) than measured (1.5–2.0%) to account for the observed changes. Our principal conclusions are similar to those of Elliott [11], Yang et al. [12], and Zhou et al. [10]. We agree that, although some local configurations may contribute to PD, the phenomenon cannot be explained in terms of these local changes alone.

The most significant photostructural change observed using the Raman scattering technique is in the region of the low-frequency or Boson peak at a wavelength of about $25\,cm^{-1}$. This Raman peak is accounted for as signifying certain medium-range order in the amorphous structure [13–15]. The position and the intensity of the peak can be used to estimate the structural correlation length $R_c \approx (\omega_{max}/V)^{-1}$, where ω_{max} is the frequency of the peak maximum and V is the sound velocity, and to characterise the degree of structural ordering. The structural correlation length has generally been associated with the dimensions of some cluster-like structures. For a-As_2S_3, the R_c value is about $7.6\,\text{Å}$. The clusters are possibly in the form of linked pyramids. In spite of the absence of final conclusions on the nature of the low-energy excitations in disordered solids, it is obvious that differences between the boson peak in the annealed and photodarkened state are due to changes in medium-range order. A decrease in the peak height and a shift to higher ω-values may be attributed to an increase in structural randomness. This is consistent with an increase in structural disorder under illumination detected in extended X-ray absorption fine structure (XAFS) spectra [12, 16] and in X-ray scattering [17]. Thus our experimental results on Raman scattering agree qualitatively with the hypothesis that PD is a photostructural change and is the result of changes in medium-range order (e.g. weakly linked AsS_3 pyramids move with respect to each other as proposed in [18]). This mechanism does not involve covalent bond breaking.

Let us consider the role played by electronic gap states in photoinduced changes. We speculate that the mechanism of PD consists of the following. Bandgap radiation initially causes excitation of the electron subsystem. The incident photon creates electron–hole pairs. These photogenerated electrons and holes may be assumed to be initially in free (mobile) states in the bands. Part of the excess carrier energy is transformed into heat. In the first place, the photogenerated electrons and holes are both rapidly trapped at the defect states. These defects cause local levels in the energy gap and affect the physical properties (i.e. optical, dielectric, and electrical). We identify such gap centres as arising from charged native defects such as C_3^+ and C_1^- in pure selenium. After trapping, the lattice surrounding the defect distorts. The resulting local relaxation evidently stabilises the trapped carriers, which must then produce excess defect centres. The latter efficiently affect the photoconductivity and transport properties of amorphous chalcogenides (e.g. enhance carrier trapping as shown by our TOF and xerographic experiments). Since the metastable states are local potential minima after the relaxations, a potential barrier must exist, retarding the transition to the initial state. Annealing of the material removes the new defects and restores the original balance. Therefore, the picture that we used is that normally empty precursor states (in other words, states that exist in quasi-equilibrium before the system is illuminated) become filled during illumination. As a result of the non-equilibrium occupation of electronic gap states, the physical properties (optical and photoelectronic) are altered.

With regard to the low-temperature data, we suggest that there exists a kind of defect that is closely related to this component of PD. Indeed, the TSC truly reflects the release from some localised states. TSC and grating experiments provide strong evidence that carrier trapping at these states is the primary effect whereas the structural change (and concomitant change in the absorption coefficient and refractive index) is the secondary effect. In a-$As_{0.5}Se_{0.5}$, the corresponding activation energy is about 0.4 eV. Since the TOF measurements show that the only mobile carriers in $As_{0.5}Se_{0.5}$ are positive holes, the above trapping centres are most likely to act as hole traps.

At room temperature, deep trapping of injected carriers (of both signs) is enhanced in photodarkened samples. Such illumination can cause an increase in deeply trapped space-charge density induced by repetitive TOF cycling but has no appreciable effect on drift mobility value. These observations support the conclusion that the main influence of bandgap illumination on $As_x Se_{1-x}$ alloys is the generation of deep states, which act as efficient hole ($0.8 \leq E_i^h \leq 0.9$ eV) and electron ($0.9 \leq E_i^e \leq 1.0$ eV) traps.

It is of interest to point out that the "memory" effect (the persistence of light-induced metastabilities) becomes appreciably prolonged with increasing As content. These investigations show the importance of arsenic-related defects in photoinduced changes. Even without a detailed understanding of the microscopic mechanism, the relationship between the metastable trap lifetime and chemical composition can be interpreted in terms of As-induced deep centres. We suggest that the addition of As to a-Se increases the total number of deep traps, which are predisposed to become photodarkened states. TOF experiments, that is, the absence of electron drift as the As content becomes appreciable ($x > 0.2$), indicate that As introduces a sufficient concentration of deep electron traps to eliminate electron transport. The network modifications (induced by threefold-coordinated As atoms) and the change in the mobility gap structure may, of course, be related.

Finally, the results tempt us to assume that the observed photoenhanced deep trapping is a universal property in amorphous chalcogenides subjected to bandgap light.

References

1. J. Orenstein, M. Kastner, V. Vaninov, Philos. Mag. B **46**, 23 (1982)
2. A.E. Owen, A.P. Firth, P.J.S. Ewen, Philos. Mag. B **52**, 347 (1985)
3. S.A. Solin, G.N. Papatheodorou, Phys. Rev. B **15**, 2084 (1977)
4. R.J. Nemanich, G.A.N. Connell, T.M. Hayes, R.A. Street, Phys. Rev. B **12**, 6900 (1978)
5. N.L. Slade, R. Zallen, Solid State Commun. **30**, 357 (1979)
6. P.J.S. Ewen, M.I. Sik, A.E. Owen, Solid State Commun. **33**, 1067 (1980)
7. U. Strom, T.P. Martin, Solid State Commun. **29**, 527 (1979)
8. D.J. Treacy, P.C. Taylor, P.B. Klein, Solid State Commun. **32**, 423 (1979)
9. O.I. Shpotyk, V.N. Kornelyk, I.I. Jaskovec, Zh. Prickl. Spektrosk. **52**, 602 (1990)
10. W. Zhou, D.E. Sayers, M.A. Paesler, B. Bouchet-Fabre, D. Raoux, Phys. Rev. B **47**, 686 (1993)
11. S.R. Elliott, J. Non-Cryst. Solids **81**, 71 (1986)
12. S.Y. Yang, M.A. Paesler, D.E. Sayers, Phys. Rev. B **36**, 9160 (1987)
13. A.J. Martin, W. Brenig, Phys. Status Solidi B **63**, 163 (1974)
14. R.J. Nemanich, Phys. Rev. B **16**, 1655 (1977)
15. V.K. Malinovsky, A.P. Sokolov, Solid State Commun. **57**, 757 (1986)
16. L.F. Gladden, S.R. Elliott, G.N. Greaves, J. Non-Cryst. Solids **106**, 189 (1988)
17. K. Tanaka, Rev. Solid State Sci. **2–3**, 644 (1990)
18. G. Pfeiffer, M.A. Paesler, S.C. Agarwal, J. Non-Cryst. Solids **130**, 111 (1991)

Chapter 8
Electronic Properties of Materials with Gross Permanent Photoinduced Changes: Cu–As–Se Glasses

Fundamental physical properties of Cu–As–Se glasses along $Cu–As_2Se_3$ tie lines have been investigated in order to obtain information on the electronic structure. Progressive addition of Cu to As_2Se_3 induces a gradual red shift of the absorption edge and significantly reduces the value of dark conductivity and its activation energy. The latter is also confirmed by thermally stimulated depolarisation current experiments. Taken together, all of the data, including steady-state photoconductivity measurements, indicate that the "electrical" and "optical" band gaps gradually narrow with Cu addition. It is established that Cu addition in As–Se glasses causes a significant change in the dark conductivity value and its activation energy. Cu–As–Se glasses are photoconductive and the temperature dependence is normal, showing the expected slope in the low-temperature region below the maximum. Thermally stimulated currents were also present in these materials. On the basis of these observations, the electronic structure of Cu–As–Se is considered from the point of view of localised states.

Although the physical properties and electronic structure of elemental Se and Se-based binary glasses have been investigated extensively, little is known about the electronic structure of ternary chalcogenides. This is especially true for the case of Cu alloyed with As_2Se_3; knowledge of the fundamental properties of ternary Cu–As–Se glasses is fragmentary [1–10]. At the same time, Cu–As–Se glasses can be used for optical data storage and in various technical devices [11].

In the present chapter, some fundamental investigations (e.g. dc conductivity, photoconductivity, thermally stimulated currents (TSCs) and direct structural probing by X-ray diffraction (XRD)) are reported to improve our knowledge of the electronic structure of $Cu–As_2Se_3$ glasses. Based on the experimental results, we propose a simple model for the influence of copper addition on the electronic states localised in the bandgap of As_2Se_3.

Bulk samples were made by mixing pure elements and loading them into clean quartz ampoules (inner diameter 8 mm). The ampoules were then sealed under a vacuum of 10^{-6} Torr. The $Cu_x(As_xSe_{1-x})_{100-x}$ ingots were kept at 950°C for about 80 h in a rocking furnace, and then cooled down to 850°C, followed by quenching in air (compositions $0 \leq x \leq 30$ at%) to room temperature. Note that these compositions were from the glass-forming region reported by Borisova [9]. Powder XRD spectra were taken to ensure that there were no crystalline inclusions in the samples.

V.I. Mikla and V.V. Mikla, *Metastable States in Amorphous Chalcogenide Semiconductors*, Springer Series in Materials Science 128,
DOI 10.1007/978-3-642-02745-1_8, © Springer-Verlag Berlin Heidelberg 2010

The bulk samples were obtained in the usual manner and then polished with diamond powder and pastes. For measurements of dc conductivity, the samples were platelets of typically $2 \times 2 \times 0.5\,\text{mm}^3$ size. As electrodes, vacuum thermally evaporated gold electrodes were used. Experimental details of dc conductivity, photoconductivity and other electrical measurements can be found in our earlier publication [4].

The optical absorption coefficients of the glasses were obtained by measuring the spectral dependence of the optical transmittance T_{op}. The sample thickness d was varied from $100\,\mu\text{m}$ to $1\,\text{mm}$. The absorption coefficient α was calculated using

$$T_{\text{op}} = (1 - R)^2 \exp\left(-\alpha\,d\right), \tag{8.1}$$

where R is the reflectivity, which is evaluated from the transmittance ($\sim 60\%$) in the near-infrared region.

For Urbach tails to be investigated quantitatively, the absorption coefficient α in these regions is described by an exponential function, $\alpha \propto \exp\left(\hbar\omega/kT_{\text{u}}\right)$, where T_{u} is the characteristic temperature in the Urbach region.

8.1 Structure

Typical XRD patterns in $Cu_x\,(As_x Se_{1-x})_{100-x}$ glasses are shown in Fig. 8.1. We start with As_2Se_3. Three halo peaks are clearly seen in the patterns of As_2Se_3, namely the characteristic first sharp diffraction peak (FSDP) located at $2\theta \approx 17°$, which is attributed to medium range structural order [12], and the second and the third peaks, located at $2\theta \approx 30°$ and $52°$, respectively. The latter peaks are due to the correlations between the second nearest neighbors and between the nearest

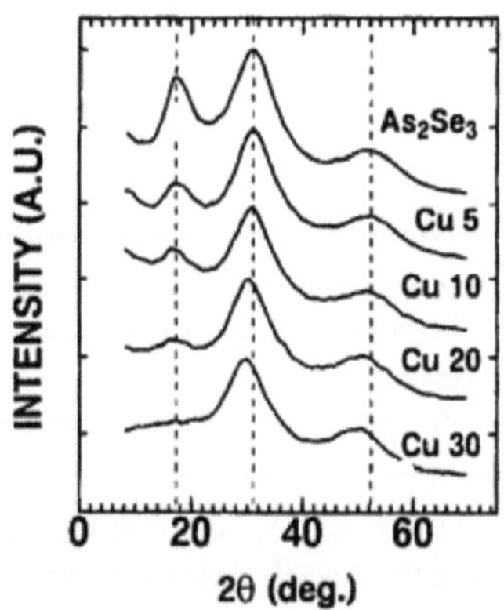

Fig. 8.1 X-ray diffraction patterns obtained for Cu–As–Se glasses. The intensities are normalised by the peaks at $2\theta \approx \sim 30°$. Cu content is shown for each curve

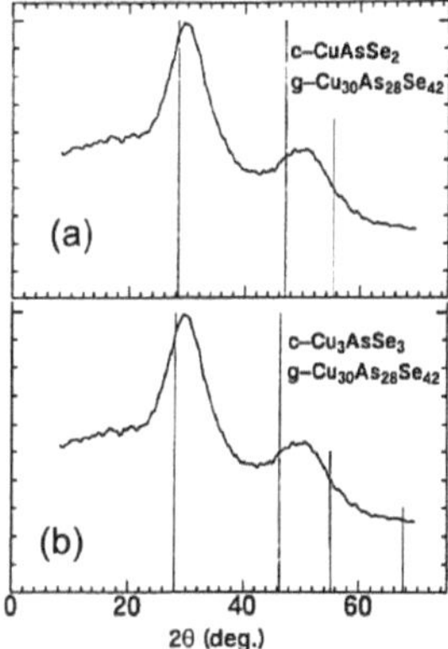

Fig. 8.2 X-ray diffraction patterns of glassy and crystalline Cu–As–Se

neighbors. On alloying As_2Se_3 with Cu, we observe the following effects: (a) the FSDP weakens and even disappears at $x > 15$; (b) additionally, the peaks at $2\theta \approx 30°$ and $52°$ shift to lower angles when Cu content increases. Here we note that these changes are in accordance with studies by Liang et al. [7] and Itoh [5].

It is well established that amorphous chalcogenide semiconductors have short-range order very close to their crystalline analogues [7, 13]. Thus, it seems reasonable to compare their diffraction patterns. Figure 8.2 shows the result of such a comparison between the diffraction patterns of g–As_2Se_3 and g–$Cu_{30}As_{28}Se_{42}$ with powder diffraction patterns of their crystalline analogues c–As_2Se_3, c–$CuAsSe_2$ and c–Cu_3AsSe_3.

As can be clearly seen, all three of the halo peaks in g–As_2Se_3 are located at positions where intense diffraction peaks in c–As_2Se_3 are observed (this experimental fact was reported first by Vaipolin and Porai-Koshits [10]). In addition, one can see that the positions of the halo peaks in Cu–As–Se ternary glasses are close to diffraction peaks in the Cu–As–Se crystals [7, 13]. It is important to note that Cu–As–Se alloys, being in crystalline or noncrystalline forms, exhibit no diffraction peak at $2\theta \approx 17°$, which is the FSDP of glassy As_2Se_3.

Summarising the above diffraction experimental results and taking into account diffraction data on the corresponding crystals, we may deduce remarkable features of Cu–As–Se glasses structure depending on the Cu content: (a) Cu–As–Se ternary glasses with Cu content less then 15 at% exhibit three halo diffraction peaks similar to that in As_2Se_3, and this means the preservation of short- and medium-range order; (b) at the same time, for $x > 15$ at%, two halo peaks in the diffraction patterns of Cu–As–Se glasses undoubtedly indicate the presence of crystalline phase in short-range order; (c) structural units at short- and medium-range intervals differ for $x < 15$ at% and $x > 15$ at%.

8.2 Electronic Properties

The temperature dependence of dc (dark) conductivity for $Cu_x (As_x Se_{1-x})_{100-x}$ glasses plotted as $I(T) = f(10^3/T)$ is shown in Fig. 8.3. It is clearly seen that the dc conductivity σ can be written in the form

$$\sigma = \sigma_0 \exp(-E_\sigma / kT). \tag{8.2}$$

Here σ_0 is the so-called pre-exponential factor and E_σ the activation energy. The value of σ_0 is approximately $10^2 \, \Omega^{-1} \, cm^{-1}$ and suggests conductivity determined by the transport states.

The dc conductivity σ and its activation energy E_σ in $Cu_x (As_x Se_{1-x})_{100-x}$ glasses gradually decreases with Cu content. The dependence of E_σ (together with other parameters) on copper content is compiled in Table 8.1.

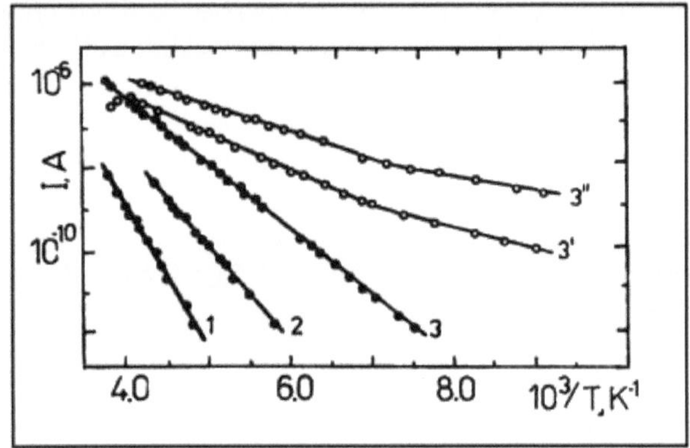

Fig. 8.3 Temperature dependence of DC dc conductivity of $Cu_x (As_2 Se_3)_{1-x}$ glasses. Cu content is 5, 10, and 15 at% (*curves* 1–3, respectively). Also shown are the temperature dependencies of photocurrent for samples under illumination with power density $0.06 \, P_{max}$ (3') and P_{max} (3'') $\left(P_{max} \approx 4 \times 10^{-2} \, W \, cm^2\right)$

Table 8.1 Activation energies of dc conductivity E_σ, thermally stimulated depolarisation current E_{TSDC}, photoconductivity E_{ph}, and trapping states depth E_t in $Cu_x (As_2 Se_3)_{1-x}$ glasses

Cu content [at%]	E_σ [eV]	E_{TSDC} [eV]	E_{ph} [eV]	E_t [eV]
0	0.90	0.90	0.31	0.40
5	0.60	0.60	0.29	0.40
10	0.45	0.41	0.24	0.36
15	0.33	0.33	0.30	0.30

8.3 Thermally Stimulated Currents

Conventional thermally stimulated currents (TSCs) and thermally stimulated depolarisation currents (TSDCs) are shown in Figs. 8.4 and 8.5.

From the slope of the initial part of corresponding curves the activation energies E_t and E_{TSDC} were determined. Here we note that E_σ and E_{TSDC} correlate, indicating that the charge carrier transitions in both cases were identical. This means, as we have shown earlier [12], that in the process of measuring TSDC one simply observes only the capacitor discharge due to dc conductivity increasing with temperature. The reader can find the detailed explanation of this problem in an excellent review by Agarwal [14]. The capacitor under consideration consists of the metallic electrodes, insulating sheets and the sample (sandwich-like configuration). At the

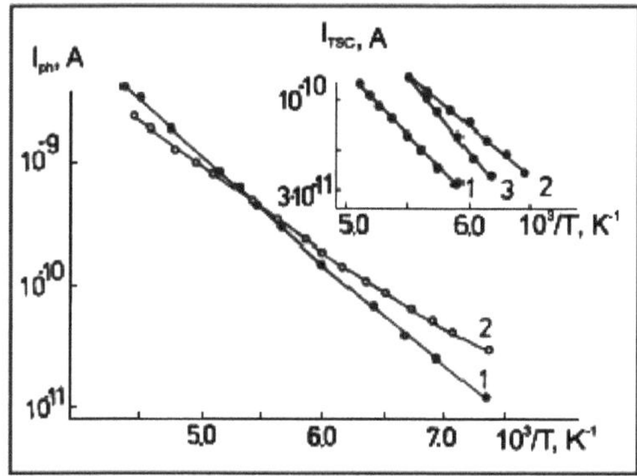

Fig. 8.4 Temperature dependencies of photocurrent for $Cu_x(As_2Se_3)_{1-x}$, $x = 0.10$ at%, at $P = P_{max}$. (1) Virgin sample, (2) sample pre-irradiated sample during for $t_i = 30$ min with P_{max} at $T = 100$ K. In the inset the initial part of TSC is shown: (1) irradiation at $T = 100$ K with P_{max}, $t_i = 1$ min; (2) irradiation at $T = 100$ K with P_{max}, $t_i = 30$ min; (3) irradiation during coolingwhile the sample is cooled from $T = 300$ K to $T = 100$ K

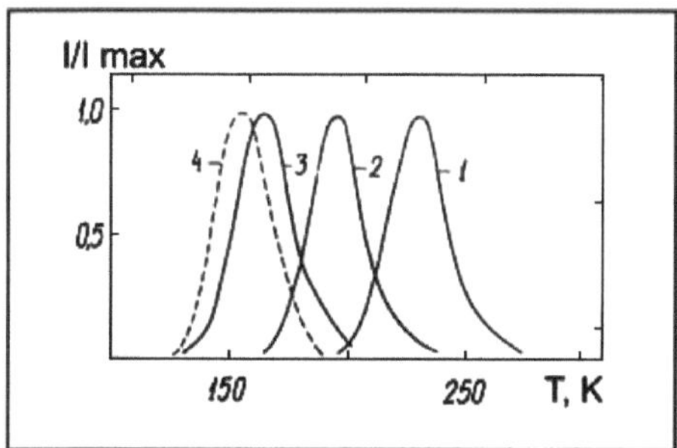

Fig. 8.5 Thermally stimulated depolarisation in $Cu_x(As_2Se_3)_{1-x}$. Cu content: 5 at% (1), 10 at% (2), 15 at% (3). Also shown is the depolarisation curve (4) for the sample photopolarised at $T = 100$ K

same time, even in this case, the TSDC can give information about $E_\sigma \approx E_{\text{TSDC}}$, especially when it is difficult to realise satisfactory contact. The TSDC maximum shifts to lower temperatures, corresponding to E_σ decreasing (Fig. 8.5). Therefore, TSDC may be used as a contact-free method for determining the conductivity activation energy E_σ. Only in the case where one is faced with filling of the traps with nonequilibrium charge carriers does the TSDC contain information on localised bandgap states.

From the study of TSC (in conventional mode and at conditions of persistent internal polarisation), it is seen that, near the valence band tail, states exist at energy E_t with respect to the mobility edge. Comparison of the corresponding activation energy values is possible for glasses containing less then 5 at% Cu – only these compositions exhibit a distinct maximum in the depolarisation curve (see Fig. 8.5). In contrast, in $Cu_x (As_2Se_3)_{1-x}$ glasses with Cu content greater than 5 at%, due to increased influence of equilibrium (dark) conductivity, difficulties arise in measuring depolarisation curves in previously photopolarised samples. In addition, we have observed some light-induced effects on electronic properties. For example, activation energies E_a determined from TSC and TSDC curves for samples illuminated by bandgap light at 100 K were found to be lower in comparison with E_t (Fig. 8.4, inset) when thermo-stimulated curves were measured with the standard procedure. Similar changes were observed when investigating the temperature dependence of the photoconductivity. The temperature dependence of the photocurrent in the glasses under study consists of three regions: (1) the pre-exponential region at low-temperature; (2) an exponential rise

$$I_{\text{ph}} = A \exp\left(-E_{\text{ph}}/kT\right), \tag{8.3}$$

where A is a constant and E_{ph} the activation energy of photoconductivity determined from the slope of the corresponding dependence; and (3) the maximum at "high" temperatures.

As expected, the absorption edge has a shape typical for chalcogenide glasses and the absorption coefficient can be described by the dependence

$$\alpha \propto \exp\left(h\omega/kT_u\right), \tag{8.4}$$

where T_u is the reciprocal of the slope of the absorption coefficient. The addition of Cu to As_2Se_3 or, in other words, alloying As_2Se_3 with Cu for the compositions studied, is accompanied by three types of effects. The first is an increase in the absorption coefficient value. Next, after an initial slight decrease in the slope of the above dependence compared to As_2Se_3 this parameter remains unchanged, independent of Cu content. That is T_u remains constant in the range 5–30 at%. Finally, we observe a gradual shift of absorption edge to lower energies with increasing Cu content. It is important to note that the rate of decrease of the activation energy E_σ is larger than that of the optical gap E_g.

In Fig. 8.6 the energy differences between the Fermi level, on the one hand, and trapping states and mobility edge, on the other hand, are shown. In the same figure the parameter δ represents the range of localised tail states. In other words, it may

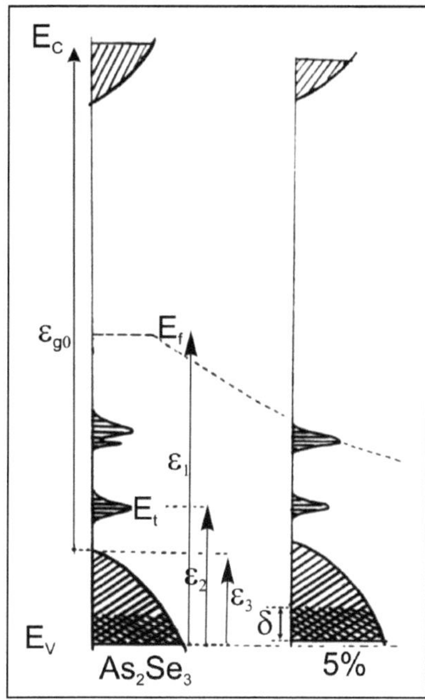

Fig. 8.6 Gap-states distribution in As_2Se_3 and $Cu_x(As_2Se_3)_{1-x}$ glasses (the latter is exemplified for by 5 at% Cu)

be regarded as a measure of the width of the valence band tail. In the Cu–As–Se system, δ falls with introduction of Cu. This indicates that Cu addition narrows the valence band tail.

Before discussing the experimental results presented, two essential facts should be emphasised: (a) the decrease in bandgap energy with increasing Cu content in the Cu–As–Se system; and (b) in the system considered, "optical" bandgap energies were greater than estimated from conductivity measurements, $E_g > 2E_\sigma$.

In order to explain the experimental results, the modified Mott–Davis model seems to be appropriate. The model is illustrated schematically in Fig. 8.6. In addition to the tails close to the mobility edges, the density of states contains several peaks. These correspond to states localised in the mobility gap. The energy interval between E_f, E_t, etc. and E_v are clearly seen in Fig. 8.6. The activation energy of the photoconductivity below the maximum in the model determines the width of the region of states localised in the tails of the corresponding bands. Another possible explanation is that E_{ph} may be attributed to some defect states. The change in E_{ph} decreases with Cu content. In terms of the model we propose, this means that the width of the tail states also decreases. It is important to note here the correlation in E_t and E_{ph} change with increasing Cu content in $Cu_x(As_2Se_3)_{1-x}$ glasses. In other words, the peak in the density of states (traps) is the same for the range of copper concentration studied. It is assumed that this peak will overlap with states localised near the Fermi level at increasing Cu concentration. Probably, this is why one can observe only one set of shallow traps in glasses with $E_\sigma < 0.7\,\text{eV}$. The influence of pre-irradiation (photoinduced changes of physical parameters) on the spectrum of localised states, we believe, is similar to the effect of Cu addition to As_2Se_3. The width of the tail states decreases in both cases. The results of TSC measurements and the temperature dependence of photoconductivity support this suggestion. Liu and Taylor [8] have found that the magnitude of the photodarkening (PD) effect decreases very rapidly with increasing Cu concentration and that the PD effect is eliminated in $Cu_x(As_2Se_3)_{1-x}$ for $x \geq 5$ at.%. However, in the present chapter we shall not consider the PD effect and the emphasis is on light-induced electronic effects.

Amorphous As_2Se_3 containing Cu exhibits a much larger value of hole drift mobility value than that of electrons (note that we cannot detect the electron transient signal even for 0.1 at% Cu). This may be a strong argument in favor of E_σ reflecting the energy difference between the valence band mobility edge and the Fermi level.

The fact that for the compositions studied E_g values are significantly greater than $2E_\sigma$ can be explained as follows: (a) copper addition shifts the Fermi level toward the valence band; (b) the mobility edge of the valence band shifts to higher energies; c) both of the above occur simultaneously, giving an "attractive" effect.

In the concentration region studied both E_σ and E_g decrease with increasing Cu content but the more significant change occurs in E_σ. Such behaviour may be explained by the appearance of acceptor-like states, which is induced by alloying.

In addition, the possibility of the valence band tail changing with Cu addition should be mentioned. For this effect to be accounted for, we adopt here the mechanism proposed by Itoh [5]. According to this mechanism, Cu atoms added to As–Se may interact with the As–Se network, since the Cu atoms possess $3d$ orbitals that effectively overlap with the Se $4p$ orbitals. Hence, if Cu–Se bonds are assumed to be stronger than As–Se in Cu–As–Se, the Cu atoms may attract the Se atoms to form optimum Cu–Se covalent configurations. Thus, structural randomness in the Cu–Se bonding configurations may decrease with increasing structural randomness in the As–Se network. This is probably why Cu introduction narrows the valence band tail. Additional argument in favor of this effect is that T_u (in the expression for the absorption coefficient), being the measure of the tail state width, decreases with Cu addition. Note that in the present chapter we consider only the width of band tail without appealing to its concrete shape, for example exponential.

We consider the band structure of Cu–As–Se glasses in the following way. Studies performed [5, 7] to clarify the local environment around Cu atoms in Cu–As–Se glasses have shown it bonding to Se atoms with coordination numbers of 2–4. It seems important to note that Cu $3d$ bands are energetically close to the Se $4p$ bands. These constitute the top of the valence band in the "host" (As_2Se_3) material. In such a case the d bands will interact with the Se $4p$ band, forming $d-p$ bonding and $d-p$ antibonding states in the upper valence region. As a result, we have the situation where the top of the valence band will shift to the Fermi level (see Fig. 8.6). Reasonably, bandgap energies as well as the energies of gap states (traps) decrease.

In the frame of this model, we have supposed that the bottom of the conduction bands in Cu–As–Se glasses consist of As–Se antibonding states.

Finally, the nature of the Cu–Se bonds in Cu–As–Se glasses may be regarding as covalent [5]. In such a case the covalent nature of the Cu–Se probably explains the absence of Cu^+ ion migration reported [15].

8.4 Carrier Transport in Cu–As–Se Amorphous Semiconductors

In the following we describe investigations of photocurrent transients in the time-of-flight (TOF) regime in order to obtain information on charge carrier transport in amorphous $Cu_x (As_2Se_3)_{1-x}$ films. Put briefly, the shape of photocurrent transients in As_2Se_3 change with Cu addition. Drift mobility–temperature data indicate that hole transport in a–$Cu_x (As_2Se_3)_{1-x}$ alloys is controlled by a set of shallow traps at 0.40 eV above the valence band edge. On the basis of these observations, the density of gap states is considered from the point of view of multiple trapping.

As mentioned, various fundamental properties (including structure) of elemental and binary chalcogenide glasses have been investigated extensively. On the other hand, the electronic properties of ternary chalcogenides have been studied only poorly. This is especially true for Cu–As–Se glasses. The latter are known to exhibit pronounced photoinduced phenomena.

Significant PD effect in Cu–As–Se glasses and higher resolution in comparison with other materials of this group, together with reversibility, makes them especially attractive objects for optical data storage. At present, despite more than three decades of intensive study of this unique phenomenon, its origin remains unclear. Reasonably, this essentially limits the possibility of materials fabrication with the required properties, which determine the photoinduced changes of fundamental physical parameters. This is especially true for the case of Cu–As–Se glasses. There is no information in the literature, to the best of our knowledge, about the drift mobility in these ternary chalcogenides. The present section deals with charge carrier transport in a–$Cu_x (As_2Se_3)_{1-x}$ amorphous semiconductors.

For the TOF experiments amorphous Cu–As–Se thin samples were used. These were prepared by flash thermal evaporation of bulk alloy Cu–As–Se from a molybdenum quasi-closed unit, using a standard vacuum system (Model VES-5) with a base pressure of 2–3 μTorr. The starting materials for pellets were Cu–As–Se glasses prepared by the usual melt-quenching technique. Substrates were glass for optical microscopic purposes with pre-deposited Au electrodes. The top electrodes were semitransparent films of evaporated gold. Sample thickness ranged from 3 to 5 μm, and was measured by a precision interference microscope (Model MII-4, accuracy not better than 0.05 μm).

The deposition rate was $\sim 1\,\mu m\,min^{-1}$. Film thickness ranged from 3 to 5 μm. Note that a shutter protection system was used at the initial and final stages to establish steady evaporation and deposition conditions. The samples are relatively homogeneous as far as their charge transport properties are concerned. Some degree of sample inhomogeneity is unavoidable in the preparation of films unless one uses flash evaporation, but this is the case for very thin films. In addition, some sample-to-sample variation in the exact compositions is also unavoidable. Because the hole lifetime is sensitive to the relative amounts of Cu and As_2Se_3, we believe that the samples were relatively homogeneous across both the film surface and the film thickness. The total injected charge was kept constant. The latter was obtained by integrating the TOF photocurrent transient. To preclude a possible contribution of thickness variations of the sandwich structure to the resulting dispersion, differ-ent regions of the semitransparent Au contact were illuminated and the obtained photocurrent waveforms compared.

The details of the conventional TOF apparatus have been described in the exten-sive articles by Spear, Pfister and Kasap [16–18], as mentioned in previous chapters. All experiments were single-shot measurements, in which laser pulses are directed at the sample and the corresponding photocurrent transient is captured on a single-event storage oscilloscope. The oscilloscope (single-event bandwidth, 0.5 µs pulse duration, S8–14 Model) captured the whole TOF signal to enable integration of the photocurrent and hence the evaluation of the total injected charge. The spread in the photocurrent tail can be precisely measured.

As usual, TOF measurements were carried out using a short light pulse from a N_2 laser (wavelength 337 nm, pulse duration 8 ns). The duration of the pulsed light was sufficiently shorter than the carrier transit time. Between the measurements samples were short-circuited and dark rested to avoid the space charge effects that can cause a nonuniform field distribution within the bulk and a distorted TOF pho-tocurrent signal. Prior to measurements the samples were rested for two weeks in the dark at room temperature. This procedure allowed us to perform TOF measure-ments on samples with stabilised structure and, accordingly, fundamental properties. It is well known that the build-up of trapped charge within the chalcogenide sam-ple during repetitive cycles in TOF measurements is a serious problem. The latter may be overcome by using a single-shot mode of operation. There is no influence of the light pulse intensity on the TOF shape. In other words, the small signal approximation is valid. When examining the reproducibility of the TOF photocur-rent shape, we observed the stability of this characteristic for successive light pulses. Undoubtedly, this means that no space charge is accumulated during the photoin-jection and transit. When dispersion entirely masked the transit time cusp (usually determined by the position of the break in the current pulse), we defined a sta-tistical transit time from a double-logarithmic plot of the algebraically decaying current. Time-of-flight experiments were performed using a conventional measuring circuit.

Figure 8.7 summarises the results of our TOF investigations. Only hole drift mobility can be measured in a–$Cu_x (As_2Se_3)_{1-x}$ by the technique outlined above, and at temperatures above 300 K a relatively well-defined transit pulse is observed. The signal contained an initial decay in the form of a spike. We may consider three possibilities for the origin of the photocurrent decay in the pretransit region. First, as the method of excitation involves the creation of a charge sheet near the top surface

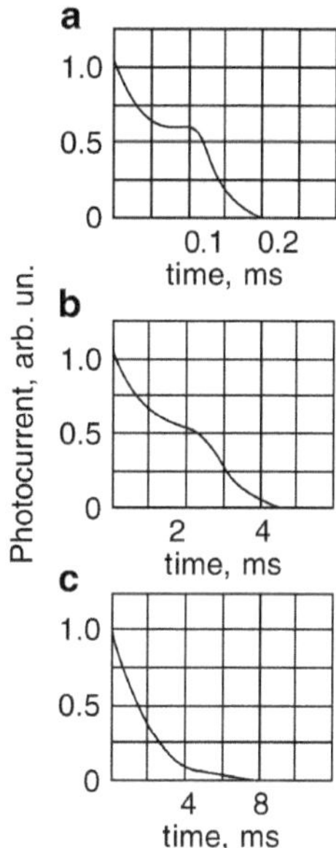

Fig. 8.7 Hole TOF photocurrent measured at $T = 325\,\mathrm{K}$ (**a**), T $= 314\,\mathrm{K}$ (**b**) and T $= 300\,\mathrm{K}$ (**c**) in $Cu_x\,(As_2Se_3)_{1-x}$ ($x = 0.05$Cu concentration 5 at%) amorphous films at applied electric field – thickness ratio $0.25\,\mathrm{V}\,\mu\mathrm{m}^{-2}$

of the sample, it has been argued that the initial spike is due to movement of holes towards the top surface. If this is the case, then, owing to the large difference in electron and hole mobility in our samples, a large spike due to holes in the electron response and a small spike due to electrons in the hole response should be observed. Furthermore, the duration of the spike should be inversely proportional to the appropriate mobility. Second, the current spike may be caused by a high-field region in the vicinity of the illuminated electrode. The initial current decay can be understood in terms of the relaxation of charge photoinjected into a distribution of localised states. A detailed analysis of transient signals in the Cu–As–Se system supports this third explanation for the spike. For samples $Cu_x\,(As_2Se_3)_{1-x}$ with $x \geq 0.05$ at any fields applied in this experiment, a shoulder is always clearly revealed in the current decays, followed by a long tail; the transit time is still discernible.

Further characteristic features should also be noted. First, and most encouragingly, TOF photocurrent shapes are very stable with respect to sample-to-sample variations. Second, for any sample we do indeed observe the true TOF signal and not any other (artefacts) – this fact is explained by experimental data for a relatively large range (5×10^4–4×10^5 V m^{-1}) of applied electric fields. The latter is illustrated in Fig. 8.8.

As expected, the transit time shortened with increasing temperature. In analogy with the behaviour observed in pure Se, hole traces for a–$Cu_x\,(As_2Se_3)_{1-x}$ show a progressive increase of transit time dispersion as the temperature is lowered

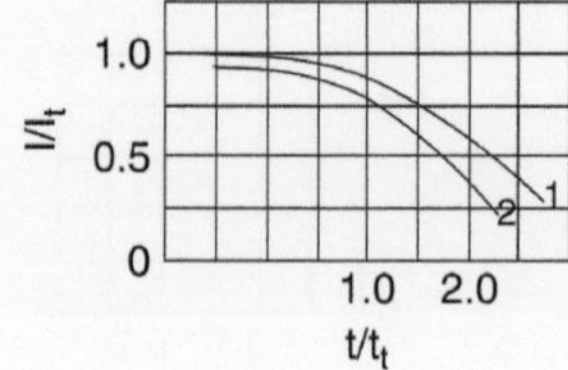

Fig. 8.8 Current traces at applied electric field $5.2\,\mathrm{V\,\mu m^{-1}}$ (1) and $4.3\,\mathrm{V\,\mu m^{-1}}$ (2). $T = 300\,\mathrm{K}$

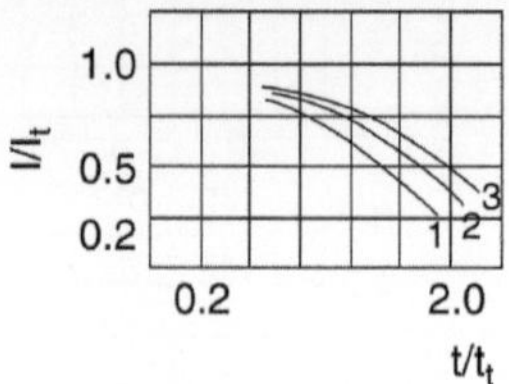

Fig. 8.9 Normalised photocurrent transients at $T = 348\,\mathrm{K}$ (1), $T = 338\,\mathrm{K}$ (2) and $T = 330\,\mathrm{K}$

(in our case from 325 K down to room temperature). It should be stressed that, as in the case of As_2Se_3, it was not possible to detect any pulses associated with the transit of electrons. Only lifetime-limited signals were observed for the case of photoinjected electrons. For these signals the transit time cannot be extracted by using even a double-logarithmic plotting. Transit pulses observed in $a\text{--}Cu_x\,(As_2Se_3)_{1-x}$ films may be characterised to first order as involving two power-law regimes:

$$I_1 \propto t^{-(1-\alpha_1)}, \tag{8.5}$$
$$I_2 \propto t^{-(1+\alpha_2)}. \tag{8.6}$$

The values of α_1 and α_2 estimated from the signal slope of the initial and final parts, respectively, are not equal to each other. As can be seen from the Fig. 8.7, the values of α_2, corresponding to the tail region of the TOF photocurrent transients, exhibit a much more rapid temperature dependence than those of α_1. It becomes clear from examination of the transients that universality of the pulse shape (predicted by a stochastic transport model) with respect to temperature, applied electric field (Figs. 8.7–8.9) and sample thickness is not observed in our case. The qualitative observation is that for transients with relatively low magnitudes, which were observed at low T, the dispersion increased as the sample temperature was reduced. The same is also valid with respect to applied field, as well as for sample thickness.

The drift mobility μ was estimated for well-defined transit time t_T according to the well-known equation $\mu = d/(Et_T)$, where d is the thickness of the sample and E is the applied electric field. Here we note that, at room temperature, the transit time scales linearly with the sample thickness, as expected for transient transport with Gaussian dispersion. Drift mobility is relatively independent of applied electric field, as actually observed in the experiments, and thermally activated with an activation energy $E_\mu \approx 0.4\,\mathrm{eV}$.

In amorphous As_2Se_3, the shape of the transient hole current even in dark-rested samples is typically dispersive and exhibits two distinct algebraic time dependences. In such a case, the transit time is identified by the intercept of the respective straight lines in the double-logarithmic plot.

Traditionally, charge carrier transport in pure and alloyed As_2Se_3 is considered within the framework of the multiple trapping models (see previous sections) and the density-of-states distribution in this material was determined from the temperature dependence of the drift mobility (Fig. 8.10) and from the post-transit photocurrent analysis. We can consider a short excitation pulse and infinitely strong absorption, so that only a thin sheet of carriers is produced in the sample near the top electrode. The injection levels are low, and we have neglected the effect of internal electric field due to the drifting carrier packet, and trap filling effects. In the present study, namely in $a–Cu_x\,(As_2Se_3)_{1-x}$, the hole transit time was measured in under the regime of equilibrium transport. This material has particularly long hole lifetimes so that deep trapping effects do not obscure the data.

The dominant feature of the density-of-states distribution in the lower half of the mobility gap is the peak at $\sim 0.4\,\mathrm{eV}$ above E_v, introduced to explain the temperature dependence of drift mobility. The presence of this peak in all of our samples, and the relative stability of its value, imply that it may be connected with intrinsic defects in $a–Cu_x\,(As_2Se_3)_{1-x}$. Note that the peak mentioned is also present in host As_2Se_3 material.

In our experiments we have observed and examined the photoinduced effects inherent to $a–Cu_x\,(As_2Se_3)_{1-x}$ amorphous layers. The influence of PD on the hole TOF transient photocurrent is illustrated in Fig. 8.11.

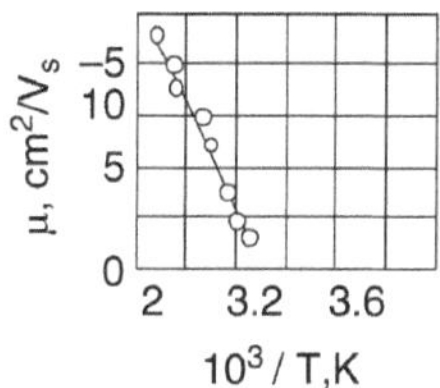

Fig. 8.10 Temperature dependence of drift mobility μ for Cu–As–Se amorphous films

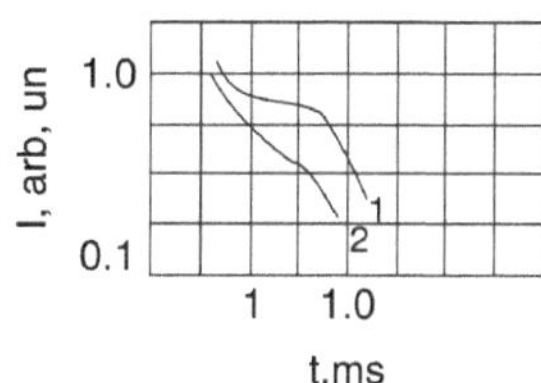

Fig. 8.11 TOF photocurrent in amorphous films either dark-rested (1) and or pre-illuminated with band-gap light (2)

Just after irradiation a drastic change in transient photocurrent was observed. As can be seen from Fig. 8.11, the dispersion of the transit pulse tends to increase with PD. In fact, the initial (pre-transit) current slope increased whereas the final (post-transit) current slope decreased. The important result of this experiment is the observation that the holes drift mobility E_μ are the same before and after light exposure. We have observed a similar behaviour of drift mobility for a number of other disordered chalcogenides. Note that, for samples photodarkened to saturation, the log I vs log t plot yielded a straight line, from which no transit time could be extracted. In such a case only a lifetime-limited signal is observed. Subsequent annealing restores the initial TOF photocurrent shape. Furthermore, the "insensitivity" of transit time to PD means that the localised states that control the transport of charge carriers and determine the activation energy of drift mobility are not responsible for photoinduced changes in the shape of photocurrent transients. Substantial changes in deep-state density, we believe, are the reason for the observed photoinduced effects. At the same time, it should be emphasised that this explanation does not exclude other possibilities.

It is obvious that bandgap light leads to the development of the pronounced variation in the space charge in the sample. The estimated amount of charge accumulated by deep traps during a ten-transit sequence doubles, at least, on PD and approaches approximately $10^{14}\,\mathrm{cm}^{-3}$. Consequently, shallow states, which control charge transport and define the activation energy E_μ [for As_2Se_3 and $Cu_x\,(As_2Se_3)_{1-x}$ nearly $0.4\,\mathrm{eV}$] of the mobility, should not undergo photoinduced changes. At the same time it should be stressed that the observed change in the current transient (namely its increased dispersion and decreased magnitude) indicate photoinduced changes of deep states with $E_t > E_\mu$. Such behaviour is probably related to enhanced carrier trapping by deep levels in photodarkened samples.

Finally, we will try to explain the effect of Cu alloying on transport properties of As_2Se_3. The influence of Cu is complicated. On one hand, the addition of Cu substantially reduces the transit time and increases the drift mobility value. On the other, it seems to decrease the magnitude of the peak in the density of states at $0.4\,\mathrm{eV}$ above E_v. A decrease in the concentration of these relatively shallow traps, in turn, causes an increase the drift mobility.

Thus, from TOF studies on amorphous $Cu_x\,(As_2Se_3)_{1-x}$ films the following conclusions have been drawn.

(1) Only hole transport is observed; in contrast, for electron transport lifetime-limited photocurrent transients were inherent.
(2) The dispersion of transit time increased with temperature lowering.
(3) Drift mobility is thermally activated with $E_\mu \sim 0.4\,\mathrm{eV}$; as for the case of As_2Se_3, the peak in the density of states located at E_μ above the valence band edge controls the hole transport.
(4) We have identified systematic photoinduced changes in hole transport with deep traps.

8.5 Concluding Remarks

Fundamental physical properties of Cu–As–Se glasses along Cu–As_2Se_3 tie lines have been investigated in order to obtain information on bandgap structure. These physical properties can be explained from the point of view that alloying As_2Se_3 with Cu leads to progressive reduction in both the "optical" and "electrical" bandgap. The latter decreases more rapidly, indicating that the valence band edge shifts to the Fermi level.

In the mobility gap of As_2Se_3 there exist a set of trapping levels located at 0.4 eV that are insensitive to Cu addition. The width of the valence band localised tail states is determined as the activation energy from the temperature dependence of the photoconductivity. Experimental results show that the width of the valence band tail states decreased with respect to that in As_2Se_3 with addition of Cu as well as with pre-irradiation of the samples.

References

1. N.F. Mott, E.A. Davis, *Electronic Processes in Non-Crystalline Materials* (Clarendon, Oxford 1979)
2. S. Neov, M. Baeva, M. Nikiforova, Phys. Status Solidi A **57**, 795 (2006)
3. M. Kitao, H. Akao, T. Ishikawa, S. Yamada, Phys. Status Solidi A **64**, 493 (2006)
4. A.A. Kikineshi, V.I. Mikla, D.G. Semak, Ukr. Fiz. Zh. **23**, 63 (1978)
5. M. Itoh, J. Non-Cryst. Solids **210**, 178 (1997)
6. M. Ohto, K. Tanaka, J. Non-Cryst. Solids **227**, 784 (1998)
7. K.S. Liang, A. Bienenstock, C.W. Bates, Phys. Rev. B **10**, 1528 (1974)
8. J.Z. Liu, P.C. Taylor, Phys. Rev. B **41**, 3163 (1990)
9. Z.U. Borisova, *Glassy Semiconductors* (Plenum, New York 1981)
10. A.A. Voipolin, E.A. Porai-Koshits, Sov. Phys. Solid State **5**, 497 (1963)
11. Y. Asahara, T. Izumitani, J. Non-Cryst. Solids **16**, 407 (1974)
12. A.A. Kikineshi, V.I. Mikla, I.P. Mikhalko, Sov. Phys. Semicond. **11**, 1010 (1977)
13. L.G. Berry (ed.), *Powder Diffraction File* (Joint Committee on Powder Diffraction Standards, Philadelphia, PA, 1974)
14. S.C. Agarwal, Phys. Rev. B **10**, 4340 (1974)
15. Y.G. Vlasov, E.A. Bychkov, Solid State Ionics **14**, 329 (1984)
16. W.E. Spear, J. Non-Cryst. Solids **1**, 197 (1969)
17. G.Pfister, H.Scher, Adv.Phys. 27, 747 (1978).
18. S.O. Kasap, in *Handbook of Imaging Materials*, 2nd edn., ed. by A.S. Diamond, D.S. Weiss (Marcel Dekker, New York 2002)

Chapter 9
Carrier Transport in Selenium-Based Amorphous Multilayer Structures

Time-of-flight experiments have been employed to determine the drift mobility of charge carriers in various selenium-based amorphous multilayer photoreceptors. The data illustrate the validity of the interpretations of the transient current signals. The addition of As to amorphous selenium progressively decreases both hole and electron drift mobility in the photogeneration layer of these photoreceptors. The results are interpreted in terms of As-induced shallow traps.

In recent years, there has been a lot of interest in amorphous Se-based multilayer photoreceptors for xerography and laser printing [1–4]. In these devices the photogeneration of charge and subsequent transport of carriers is separated functionally, that is, these processes occur in different layers. Multilayer structures are very advantageous because the composition of different layers can be tailored independently to meet specific requirements such as spectral sensitivity, good charge transport properties, stabilisation and resistance against environmental interactions. A typical double-layer photoreceptor structure consists of a relatively thin (a few micrometres) Se-based layer for carrier generation, referred to as the photogeneration layer (PGL), and a thick ($\approx 50\,\mu$m) pure selenium layer, referred to as the charge transport layer (CTL). There are also triple-layer photoreceptors, consisting of a transport layer, photogeneration layer and a third thin layer whose function is to protect the PGL layer and improve the charge acceptance of the photoreceptor.

The current trend in photoreceptor design is to use multilayer structures of various Se-based compositions. In particular, certain additives, for example As, Te, and Bi, have been shown to extend the spectral sensitivity of selenium to longer wavelengths [1,4,5]. Arsenic has been particularly effective in providing increased thermal stability and resistance to crystallisation.

In characterising the electrical properties of Se-based photoreceptors, the time-of-flight (TOF) technique [6,7] has proved to be a powerful tool in basic investigations as well as for diagnostic purposes. Typical information obtained from TOF studies are carrier mobility and number and/or depth of traps in the photoreceptor. The interpretation of TOF signals obtained from monolayer photoconductors with nondispersive transport is relatively simple and straightforward. In the case of As–Se and Se–Te alloys the carriers, which were initially localised in a narrow sheet, spread out due to trapping as they traverse the photoconductor. This causes a significant distortion of the TOF signal, which makes the evaluation much more

V.I. Mikla and V.V. Mikla, *Metastable States in Amorphous Chalcogenide Semiconductors*, Springer Series in Materials Science 128,
DOI 10.1007/978-3-642-02745-1_9, © Springer-Verlag Berlin Heidelberg 2010

difficult. Both the electron and hole drift mobility have been studied in amorphous $As_x Se_{1-x}$ alloys of different compositions (for details the reader is referred to the review articles [8,9]), but the investigations were restricted mostly to lower As concentrations. The shapes of transients have not been analyzed in more detail. Thus, we discuss the evaluation of drift mobility of charge carriers in multilayer structures and their dependence on the composition. The data show that the use of a double-layer structure is the enabling tool to obtain information on dispersive transport. The reasons for choosing a-Se for such a study in this case are the following. First, the properties of a-Se are well documented [2, 4, 6–12] and as such it can serve as an ideal test and model material for the comparison of various results on mono- and multilayer structures. Second, the exact nature of traps (both shallow and deep) in a-Se and in its technologically important alloyed forms such as $As_x Se_{1-x}$ has not been conclusively determined. A further reason for using a-Se is that it can be readily prepared by using conventional vacuum-deposition techniques with reproducible properties so that the results presented herein will be typical for any pure or alloyed a-Se film.

Single- and double-layer xerographic photoreceptors were fabricated by vacuum evaporation of vitreous pellets from stainless steel boats onto SnO_2-coated glass (or oxidised aluminium) substrates held at room temperature. Bulk $As_x Se_{1-x}$ glasses were prepared by a conventional melt-quenching method. To prepare a particular composition, appropriate quantities of high-purity constituent elements sealed in an evacuated quartz ampoule were rotated continuously to ensure thorough mixing of the constituents. The ampoules were then quenched in ice water.

The coating chamber was equipped with a set of independently controlled stainless steel boats and a shutter system to enable the fabrication of multilayer structures. Pure selenium pellets were loaded into one boat and $As_x Se_{1-x}$ alloys into another. The two sources were evaporated sequentially (without breaking the vacuum) at boat temperatures of $\sim$450 K. Typical coating rates were $1\,\mu m\,min^{-1}$. After evaporation, the samples were allowed to anneal over several weeks in the dark at room temperature. During this period, due to structural bulk relaxation, most physical properties of the photoconductor film become stabilised. The compositions of the deposited films were determined by electron probe microanalysis and the compositions quoted ($0 \leq x \leq 0.20$) are accurate to within 0.5 at%. By shuttering the beginning and the end of the evaporation, a uniform arsenic composition across the film thickness can be obtained. In all experiments, a transparent gold electrode (about $300\,\mu m$ thick) was used as the top contact.

Electroded TOF experiments have already been described previously [13]. They were carried out under small-signal conditions to determine hole and electron drift mobility. Carriers were generated by illuminating the sample through the semitransparent Au electrode with a pulsed N_2 laser at 337 nm wavelength. The duration of the pulsed light was about 10 ns, which was sufficiently shorter than the carrier transit time. Because the photocarriers are generated at the sample surface, owing to the light absorption coefficient of $As_x Se_{1-x}$ glasses at 337 nm ($>10^4\,cm^{-1}$), the species of drifting carriers can be chosen by changing the polarity of the applied voltage across the sample. The transient current was amplified and displayed on a storage oscilloscope. The transport of carriers in the corresponding layers was evaluated by time-resolving the transit signal. To avoid cumulative bulk space charge build-up, after each excitation the sample was discharged by successive nitrogen laser pulses and then allowed to rest in the dark for several minutes (so-called single-shot mode of operation).

9.1 Monolayer Systems

Figure 9.1a shows a typical TOF electron signal of a space-charge-free amorphous Se monolayer. We see that the transient profile for a well-rested (dark adapted) is a quasi-rectangular pulse. The photocurrent remains constant up to 1 μs, and then decreases abruptly. The signal indicates essentially nondispersive transport at room temperature. At the time when the carriers arrive at the substrate the signal decreases quickly to zero. This is the transit time t_T, which can be readily determined from the break point. The drift mobility is then calculated from $\mu^e = d^2 / (t_T V_a)$, where d is the sample thickness and V_a the applied voltage. This room temperature value of the drift mobility, $\mu^e \approx 5 \times 10^{-3}\,\mathrm{cm^2 V^{-1}\,s^{-1}}$, is in remarkably good agreement with reported [8, 14] values.

Addition of As to Se causes the transport to become more dispersive. Figure 9.1b shows the corresponding transient characteristic of a single layer $As_x Se_{1-x}$ structure. The signal contained an initial decay in the form of a spike. We may consider three possibilities as the origin of the photocurrent decay in the pre-transit region.

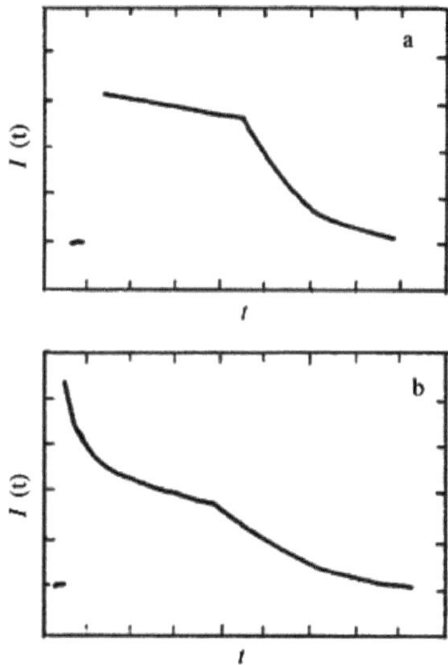

Fig. 9.1 Typical TOF traces for monolayer structures. (**a**) Pure Se, $10\,\mathrm{mA\,div^{-1}}$, $0.2\,\mathrm{\mu s\,div^{-1}}$, $9 \times 10^4\,\mathrm{V\,cm^{-1}}$, $d = 3.2\,\mathrm{\mu m}$. (**b**) $As_{0.08}Se_{0.92}$, $5\,\mathrm{\mu A\,div^{-1}}$, $20\,\mathrm{\mu s\,div^{-1}}$, $1.4 \times 10^5\,\mathrm{V\,cm^{-1}}$, $d = 3.3\,\mathrm{\mu m}$

First, as the method of excitation involves the creation of a charge sheet near the top surface of the sample, it has been argued [15] that the initial spike is due to the movement of holes towards the top surface. If this is the case, then, due to the larger difference in electron and hole mobility in our samples, a large spike due to holes should be observed in the electron response, and vice versa. Furthermore, the duration of the spike should be inversely proportional to the appropriate mobility. Second, the current spike could be caused by a high-field region in the vicinity of the illuminated electrode. Next, the initial current decay can be understood in terms of relaxation of charge photoinjected into a distribution of localised states. A detailed analysis of transient signals in the $As_x Se_{1-x}$ system supports this third explanation of the spike. For samples $As_x Se_{1-x}$ with x < 0.10 at any fields applied in this experiment, a shoulder is always clearly revealed in the current decays followed by a long tail; the transit time is still discernible. As arsenic concentration increases further the pronounced break point becomes progressively more difficult to identify. Transit pulse shapes have become highly dispersive, meaning that there is now a broad statistical distribution in photoinjected carrier transit times. In these alloys, photoinjected carrier transport is generally interpreted [4, 8] as shallow trap-controlled transport, which involves multiple trapping of transiting carriers from the band into shallow states. Thus, the increasing dispersion of TOF transients can be understood in terms of a composition-induced broadening in the respective interactive shallow trap manifolds.

9.2 Multilayer Systems

The transients corresponding to carrier movement across multilayer samples have different profiles from those from monolayer samples. A step-like transient current waveform is expected, owing to the difference in mobility between the two layers (i.e. PGL and CTL). However, as is evident from Fig. 9.2, in actual photoreceptors this is not the case, rounded waveforms are observed. The signal consists of two parts. The initial part represents the charge movement through the $As_x Se_{1-x}$ photogeneration layer.

Because of the low mobility in this material the carriers move with relatively low speed. The magnitude of the signal is small and decreases even further as carriers are trapped in the layer. At time t_0 the fastest carriers arrive at the interface between the PGL and CTL and enter the CTL. There they move with a much higher speed, because of their large mobility in Se, and induce a much larger signal in the measuring circuit. Note that the portion corresponding to transport in a-Se CTL is now distorted in comparison with that observed for the a-Se single layer. The reason for this is the spreading of the propagating carrier packet in the PGL. The signal increases with increasing number of carriers entering the transport layer. After the transit time through the CTL the carriers recombine with the charge at the substrate. The signal becomes zero when the total charge has finished the transit through the transport layer. Thus, the separate contributions from transport in the two layers are

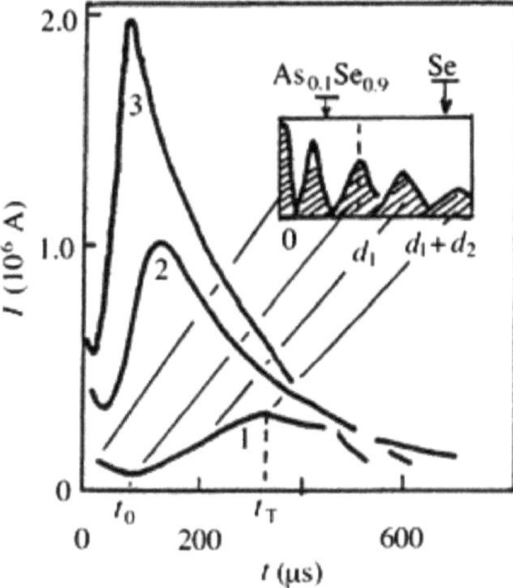

Fig. 9.2 Traces of transient electron currents in an a-As$_{0.10}$Se$_{0.90}$/Se double-layer as a function of the layer thickness of As$_{0.10}$Se$_{0.90}$: $d_1 = 0.9$ (1), 0.7 (2) and 0.5 μm (3). The applied electric field is 7.2×10^4 V cm^{-1}. The *inset* shows schematically the spatial distribution of photocarriers at different times

clearly in evidence. From the inflection points, the beginning and end of the transit time, t_0 and t_T, were measured.

The assignment of the inflection points in the waveform to t_0 and t_T is supported by the following arguments. First, t_0 and t_T scaled correctly with the thickness of PGL and CTL in TOF experiments. Next, from the knowledge of transit times t_0 and t_T, applied voltage V_0, thicknesses d_1 and d$_2$, the drift mobilities of the charge carriers μ_1 and μ_2 in the two layers are calculated from [16]

$$\mu_1 \approx \frac{d_1^2 \left(1 - \varepsilon_1/\varepsilon_2\right) + d_1 d \varepsilon_1/\varepsilon_2}{t_0 V_0}, \tag{9.1}$$

$$\mu_2 \approx \frac{d_2^2 \left(1 - \varepsilon_2/\varepsilon_1\right) + d_2 d \varepsilon_2/\varepsilon_1}{t_T V_0}, \tag{9.2}$$

where t_0 and t_T are transit times for the PGL and CTL, respectively, and $d = d_1 + d_2$ is the total thickness. Note that these data are in general agreement with those obtained for monolayer samples of the same composition and thus support the assignment of t_0 and t_T as transit times. Furthermore, t_0 and t_T are independent of the substrate materials and top contacts.

In order to examine the contact effects on both μ_1 and μ_2, TOF was carried out on similar samples with different top and bottom contacts (gold, aluminium, copper and SnO$_2$). From the results, it can be concluded that both μ_1 and μ_2 are independent of the top and bottom contacts and also of the applied field. These observations indicate clearly that μ_1 and μ_2 are meaningful parameters characteristic of bulk transport in the PGL and CTL.

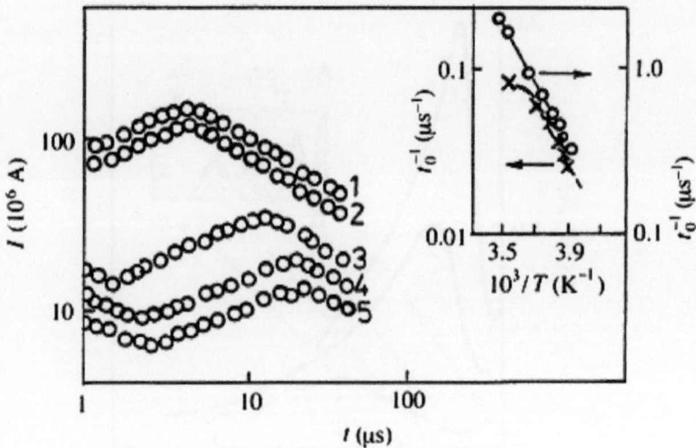

Fig. 9.3 Temperature dependence of transient electron currents in an a-As$_{0.10}$Se$_{0.90}$/Se double layer ($d_1 = 0.3\,\mu$m, $d_2 = 5\,\mu$m) at 8×10^4 V cm^{-1}. $T = 290$ (1), 283 (2), 270 (3), 265 (4), and 260 K (5). In the *inset*, the inverse of electron transit times t_0 and t_T is plotted against reciprocal temperature

Figure 9.3 shows a series of waveforms of transient photocurrent for various temperatures. We see that the transit signal does not change appreciably with temperature. Also note that at the final stage ($t > t_T$), the TOF signal exhibits a power-law decay and appears to be dispersive. According to the Scher–Montroll theory [17], in the case of a dispersive transport process, $I(t)$ should exhibit power-law dependences $t^{-(1-\alpha)}$ and $t^{-(1+\alpha)}$ for $t < t_T$ and $t > t_T$, respectively, where α is the disorder parameter. The latter can be determined from both the initial (α_i) and final (α_f) slopes. In our results the condition $\alpha_i = \alpha_f$ predicted by Scher and Montroll is generally not observed. The field and temperature dependences of α_i and α_f are different. The inset in Fig. 9.3 shows the temperature dependence of the drift parameters t_0 and t_T. The Arrhenius plot gives a good straight line in the temperature range 250–300 K, showing an activation-type process of electron transport in the present samples. The activation energy can be obtained from the slope. The values of E_μ in Fig. 9.3 are estimated to be 0.33 and 0.35 eV, which are close to those of pure Se and As$_{0.10}$Se$_{0.90}$, respectively. The results above are taken as evidence that electron transport in multilayer structures is still nondispersive.

The triple-layer device shows a transient characteristic with a more complicated signature (Fig. 9.4). The photocurrent can roughly be divided into three time regions. Transport in the thin top layer (a-Se) appears as a narrow pulse. An initial current rises to a peak value and then decays to a smaller value. This is defined as a "spike". There follows a region of low signal amplitude (the "saddle region"), corresponding to transport in the thin low-mobility middle layer. Finally, the growth of current terminates in a broad rounded peak, which is then followed by a slowly decaying tail. This broad portion of the signal ($t > t_2$) represents transport in the high-mobility bottom layer. As the applied field across the multilayer structure increases, t_1, t_2 and t_3 decrease.

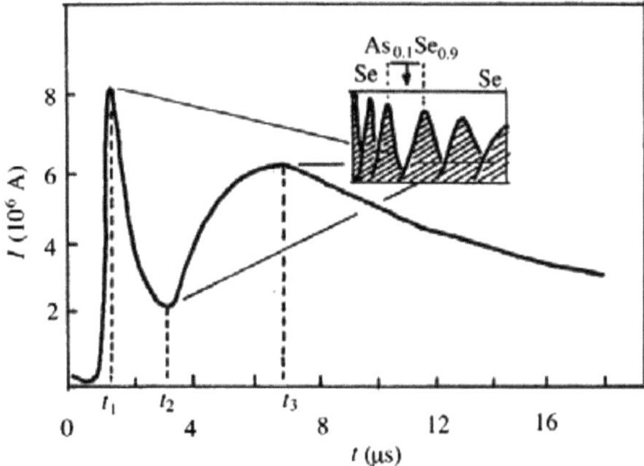

Fig. 9.4 A typical transient electron current waveform in an a-Se/As$_{0.10}$Se$_{0.90}$/Se triple-layer sample ($d_1 = 0.81\,\mu$m, $d_2 = 0.43\,\mu$m, $d_3 = 8.40\,\mu$m, $E = 8 \times 10^4$ V cm^{-1}). The inset shows the spatial distribution of photocarriers at different times

The assignment of t_1, t_2 and t_3 (see inflection points in Fig. 9.4) to transit times in the top, middle and bottom layers, respectively, is supported by the fact that the drift mobility of charge carriers for the three layers were calculated to be similar to the corresponding single layers. The general features of current waveforms described above are common to both hole and electron response.

9.3 The Effect of the Interface

A potential barrier may be formed at the interface between the As$_x$Se$_{1-x}$ and Se layers owing to the difference in the optical band gaps. Thus the transport can be influenced by injection characteristics. We have calculated the carrier injection efficiency for As$_x$Se$_{1-x}$ to a Se layer. For example, for double-layer photoreceptors, this is achieved by calculating the charge that reaches the interface between PGL and CTL and the total charge collected at the end of the transit time. Such calculations can be easily made by integrating the transient currents for each layer. The injection efficiency is independent of the applied electric field in the range 4×10^4 to 1×10^5V cm^{-1}. These results suggest that the charge accumulation at the interface does not significantly affect the charge transport. The physics of the interface states between different mobility gap amorphous layers have not been investigated in detail, and as a result it is not possible to extend the experimental observation here to amorphous semiconductors in general.

9.4 Light-Induced Effects on Photocurrent Transients

It is well known that amorphous chalcogenide semiconducting materials exhibit many kinds of photoinduced phenomena (photodarkening and photocrystallisation, to mention just two), on which extensive work is now being carried out because of their fundamental interest [18, 19]. Furthermore, the phenomena have received much attention also because of their potential application in optical memories [20]. While there are many experimental studies on the changes in structural and electronic properties, only a few studies concerning light-induced effects on photocurrent transients have been reported [21–23]. To obtain further insight into the photoinduced electronic effects, we have examined for the first time transient drift characteristics in amorphous multilayers that were exposed to bandgap light. Figure 9.5 shows a typical trace of the transient electron in a dark-rested and pre-exposed $As_{0.10}Se_{0.90}/Se$ sample. It is apparent that bandgap irradiation of the sample leads essentially to a decrease of the current level, whereas the transit time t_0 remains unchanged ($t_0 \approx 20\,\mu s$ for the data in Fig. 9.5). If, after pre-illumination, the sample is allowed to relax in the dark, the previous (dark-rested) magnitude and shape of the transient current is fully recovered. The important result of this experiment is the observation that the electron drift mobility μ^e is the same before and after light exposure. Consequently, shallow states, which control charge transport and define the activation energy E_μ^e of the mobility, should not undergo photoinduced changes. At the

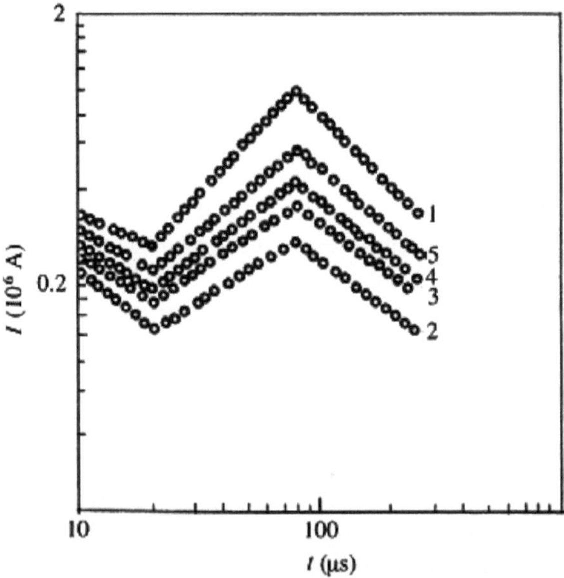

Fig. 9.5 Transient electron current in a-$As_{0.10}Se_{0.90}/Se$ double-layer samples ($d_1 = 0.5\,\mu m$, $d_2 = 9.6\,\mu m$, $E = 7.2 \times 10^4\,V\,cm^{-1}$, $T = 293\,K$): (1) dark rested; (2–5) exposed to bandgap illumination and then dark-rested for 5, 20, 30, and 40 min, respectively

same time it should be stressed that the observed change in the basic xerographic parameters, that is the dark discharge rate, initial charging potential, residual potential and its dark-decay rate, indicate photoinduced changes of deep states with $E_i > E_\mu^e$.

9.5 Use of Multilayer Structures for the Determination of Transit Time

A primary goal of TOF experiments is to extract basic material parameters that may be used for more general applications. Of particular importance are estimates of the transit times and carrier mobility. It is well known that TOF measurements of the drift mobility in As_2Se_3, As_2S_3 and some polymeric films are complicated [17, 24, 25] by a significant spreading of the carrier packet as it propagates through the sample. If carrier transport is very dispersive, a current pulse may appear featureless. However, if the current pulse is displayed on logarithmic axes, a discontinuity of gradient is observed, corresponding to the transit time of the fastest carriers.

In this section we propose that the transit time of transport photocarriers can be obtained by an analysis of transient current in a double layer that consists of the thin chalcogenide under test and another material with higher mobility, such as a-Se. Figure 9.6a shows the type of carrier transit pulse that is frequently encountered in the study of amorphous semiconductors.

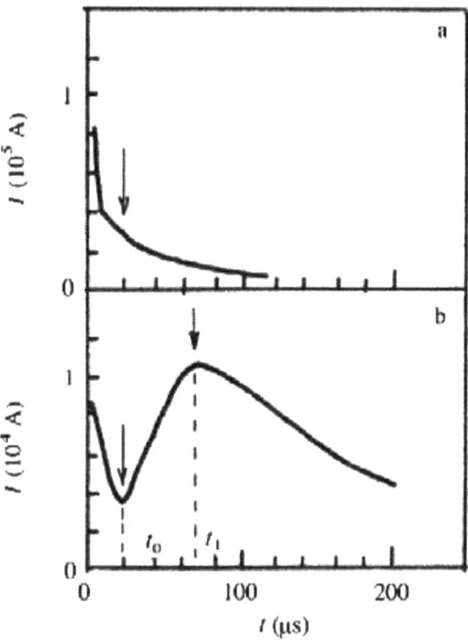

Fig. 9.6 Traces of transient hole currents in (**a**) an a-$As_{0.4}Se_{0.6}$ monolayer and (**b**) an a-$As_{0.4}Se_{0.6}$/Se double layer ($d_1 = 1.5\,\mu$m, $d_2 = 25\,\mu$m, $E = 3 \times 10^4$ V cm^{-1}

An anomalously high degree of dispersion is immediately evident from the figure. There is no "flat part", characteristic of idealised current response, to specify a transit time. The transit time is defined by log I vs log t. Note that there exist some disadvantages in using logarithmic conversion of the transient current to extract the transit time. Although the intersection of the power laws $t^{-(1-\alpha)}$ and $t^{-(1+\alpha)}$ has been used extensively to define the transit time, there is a good deal of arbitrariness to this definition. Experimentally one has access to only limited sections of pre- and post-transit currents.

On the other hand, better time resolution has been obtained in TOF experiments with multilayer structures. In the case of double-layer configuration (Fig. 9.6b), the transit time for the $As_{0.4}Se_{0.6}$ layer could be accurately assessed. The waveform clearly showed a discontinuity in the form of an inflection point at $t = t_0$. The time designated as t_1 (marked with an arrow) is used as a transit time. This technique of time-resolved transient current measurements in multilayer structures is especially suited to low-mobility, high-resistivity amorphous chalcogenides with dispersive transport characteristics.

9.6 Concluding Remarks

We have considered transient drift characteristics in Se-based multilayer devices using the TOF technique. To illustrate the method of analysis of TOF transient signals, various experimental current waveforms were presented. The shape of the photocurrent waveform indicates two separate transit times for the two layers of a double-layer structure. The gradual addition of arsenic progressively decreases μ^e. The principal observations are then an increase in dispersion with a reduction in mobility as As is alloyed with a-Se. It is reasonable, within the shallow-trap-controlled mobility model [7, 8] to interpret our observations in the following manner. In a-Se the electron transport is controlled by a narrow manifold of traps displaced $\approx 0.33\,eV$ from the conduction band mobility edge. The transport tends to become dispersive when kT is less than the nominal trap manifold width δE. Note that in the simulation studies of Marshall [25], various distributions of trapping centres were examined and it was established that, in each case, highly dispersive transit pulses could be generated if the localised states extended over more than a few kT. A distribution of the Gaussian form is capable of generating values of α_i and α_f with temperature dependence very similar to that of typical experimental data for a-Se. Addition of As to a-Se, we believe, broadens the distribution of shallow traps, thus shifting the threshold for nondispersive transport to higher temperatures. The decrease of electron mobility with increasing As content could be accounted for by an increase in the density of shallow traps, with E_μ remaining constant at approximately 0.33 eV. Unlike the electron case, the effect of As on hole transport cannot be explained by a gradual change in the density and energy distribution of shallow traps. We can attribute the lifetime-limited hole signal in the concentration range 2–6 at% As to the loss of carriers from the charge packet due to deep trapping.

These results are in good agreement with our previous data on monolayers [26] and the data of other investigators [4, 8, 27].

Further, we have shown that the transient photocurrents in double-layer structures can conveniently be used for spectroscopic purposes in situations where the use of the conventional TOF technique is limited.

Our experimental results show that, irrespective of the exact nature of interface, transport of charge carriers through double-layer samples does not experience significant deep trapping.

References

1. L. Cheung, G.M. Foley, P. Fournia, B.E. Springett, Photogr. Sci. Eng. **26**, 245 (1982)
2. C. Juhasz, M. Vaezi-Nejad, S.O. Kasap, J. Mater. Sci. **22**, 2569 (1987)
3. S. Imamura, Y. Kanemitsu, M. Saito, H. Sugimoto: J. Non-Cryst. Solids **114**, 121 (1989)
4. S.O. Kasap, in *Handbook of Imaging Materials*, ed. by A.S. Diamond (Marcel Dekker, New York 1991), p. 329
5. B.E. Springett, In *Proc. 4th Int. Symp. on Uses of Selenium and Tellurium* (Selenium-Tellurium Development Association, New York 1989), p. 126
6. W.E. Spear: J. Non-Cryst. Solids **1**, 197 (1969)
7. J.M. Marshall, A.E. Owen, Phys. Status Solidi A **12**, 181 (1972)
8. A.E. Owen, W.E. Spear, Phys. Chem. Glasses **17**, 174 (1976)
9. R.C. Enck, G. Pfister, in *Photoconductivity and Related Phenomena*, ed. by J. Mort, D.M. Pai (Elsevier, New York 1976)
10. M. Abkowitz, R.C. Enck, Phys. Rev. B **25**, 2567 (1982)
11. G. Juska, J. Non-Cryst. Solids **137/138**, 401 (1991)
12. G. Juska, K. Arlauskas, E. Montrimas, J. Non-Cryst. Solids **97/98**, 559 (1987)
13. V.I. Mikla, Ph.D. Thesis, (Odessa State University 1984)
14. M.D. Tabak, P.J. Warter, Phys. Rev. **173**, 899 (1968)
15. D.J. Gibbons, A.C. Papadakis, J. Phys. Chem. Solids **29**, 115 (1968)
16. S.M. Vaezi-Nejad, Int. J. Electron. **62**, 361 (1987)
17. H. Scher, E.W. Montroll, Phys. Rev. B **12**, 2455 (1975)
18. K. Tanaka, Rev. Solid State Sci. **2/3**, 644 (1990)
19. S.R. Elliott, J. Non-Cryst. Solids **81**, 71 (1986)
20. K. Tanaka, MRS Int. Mtg. Adv. Mater. **12**, 225 (1989)
21. V.I. Mikla, D.G. Semak, A.A. Kikineshi, Ukr. Fiz. Zh. **25**, 2021 (1980)
22. L.P. Kasakova, L. Toth, M.A. Taguyrdzhanov, Phys. Status Solidi A **71**, K 107 (1982)
23. M. Itoh, K. Tanaka, Jpn. J. Appl. Phys. **34**, 2216 (1995)
24. G. Pfister, H. Scher, Adv. Phys. **27**, 747 (1978)
25. J.M. Marshall, Rep. Prog. Phys. **46**, 1235 (1983)
26. V.I. Mikla, D.G. Semak, A.V. Mateleshko, A.A. Baganich, Phys. Status Solidi A **117**, 242 (1990)
27. J. Schotmiller, M. Tabak, G. Lucovsky, A. Ward, J. Non-Cryst. Solids **4**, 80 (1970)

Chapter 10
Spectroscopic Studies of Gap States and Laser-Induced Structural Transformations in Selenium-Based Arsenic-Free Amorphous Semiconductors: Sb_xSe_{1-x} Alloys

Amorphous Sb_xSe_{1-x} films have been prepared by thermal evaporation in a vacuum at 10^{-6} Torr and their electrical properties were examined. A combination of xerographic and time-of-flight (TOF) techniques was employed to study the gap states. Xerographic dark discharge experiments on a-Sb_xSe_{1-x} films indicated that the decay of the surface potential over the time scale of observation ($\sim$600 s) is essentially due to bulk thermal generation of electrons, and their subsequent sweep-out and depletion. Electron emission occurs from midgap-localised states. When a-Se is alloyed with Sb the dark discharge becomes more rapid, owing to an increase in the volume density of the midgap electron emission centres with Sb concentration. The repetition of the xerographic cycle many times leads to saturation of the surface residual voltage, which was used to determine the concentration N_t of deep electron traps. These cycled-up xerographic residual voltage measurements indicate that the saturated residual voltage increases with Sb addition, owing to an increase in the concentration of electron deep traps $\left(N_t \sim 10^{15}\,cm^{-3}\right)$. The xerographic photosensitivity for Sb_xSe_{1-x} alloys is greater at longer wavelengths ($\lambda > 670\,nm$) than for pure Se. Time-of-flight experiments indicate that electron transport in a-Sb_xSe_{1-x} alloys is controlled by a set of shallow traps at $\sim$0.33 eV below E_c whose concentration increases with Sb addition. Laser-induced transitory changes in the optical properties and a phase transformation to a crystalline modification (photocrystallisation) were considered. The photoeffects observed critically depend on exposure (intensity) and exhibit threshold behaviour. When the irradiation intensity is more than the threshold value, photoinduced crystallisation takes place. The origin of this phenomenon is discussed on the basis of the microcrystalline model.

The physical properties of chalcogenide glasses that are arsenic free have attracted much interest for physical and technical reasons. Among these non-crystalline materials, the Sb_xSe_{1-x} glasses are attractive candidates for applications requiring low melting temperatures, low thermal conductance and high viscosity. In particular, thin films of the Sb–Se system, which can undergo an amorphous–crystalline phase transition, have been studied as candidate materials for reversible optical recording.

Chalcogenide glassy semiconductors, which are based on sulfides, selenides and tellurides of the main group III–V elements, have recently gained much interest as

V.I. Mikla and V.V. Mikla, *Metastable States in Amorphous Chalcogenide Semiconductors*, Springer Series in Materials Science 128, DOI 10.1007/978-3-642-02745-1_10, © Springer-Verlag Berlin Heidelberg 2010

materials for infrared-transparent optical fibres, photoconductors, photoresists, optical memories, optoelectronic circuits, etc. [1–4]. In general, glasses are often preferred over crystalline compounds with similar characteristics because of favourable mechanical and interfacing properties (i.e. properties that can be influenced by the presence of interfaces between various layers in multilayer structures and heterojunctions) and good processability. Further, a lack of translational regularity often makes it possible to tailor physical properties to specific applications by adjusting chemical composition. Selenium in the non-crystalline phase is particularly attractive in semiconductor physics research not only because of its commercial importance as a xerographic photoreceptor material but also because of its interesting physical properties [2–4]. The elemental nature, relative structural simplicity, ambipolarity, high photosensitivity, etc. make Se a particularly suitable object for the study of the effect of impurities and/or alloying on the electrical properties. The alloying effects in a-Se have been well documented and reviewed for such additives as As, S, Te, P, see for example [3, 5–8]. At the same time, there is only limited information about the basic properties of the amorphous Sb–Se system [9–11]. Recently, thin films of these materials have become of interest for use in data storage [10]. In the following we report the effect of antimony on the deep and shallow states of selenium. These states are the thermal emission (trapping) centres that control the slow dark decay, and the first and cycled-up residual surface potential on capacitively charged specimen films.

Glassy $Sb_x Se_{1-x}$ alloys ($x < 0.05$) were prepared by conventional melt quenching. Cleaned silica tubes containing a mixture of the appropriate amount of constituents Sb and Se were evacuated to 10^{-5} Torr and sealed. The contents of the tubes were melted in a furnace and continuously agitated for 10 h to ensure good homogeneity. The melt was rapidly quenched in cold water from 800 K and the cooling rate was estimated to be $200\,K\,s^{-1}$.

Amorphous films were prepared by thermal evaporation of the $Sb_x Se_{1-x}$ source material from an open stainless steel boat onto pre-oxidised Al substrates in a vacuum of $\sim 10^{-6}$ Torr. The substrate was cleaned in neutral detergent, de-ionised water, ethyl alcohol and acetone and then oxidised in air at $1,800°C$ before deposition. Deposition of the $Sb_x Se_{1-x}$ alloys was performed at a substrate temperature of 300 K. Note that while the substrate is aluminium, the layer directly under the film is aluminium oxide. Care was taken to avoid Sb fractionation. We point out that, by excluding the initial and final stages of the evaporation process by shuttering, the deposited antimony content is relatively uniform across the film thickness, as can be seen from our electron probe microanalysis data, and close (± 0.5 at%) to the starting source value. A typical coating rate was $1–2\,\mu m\,min^{-1}$. The film thicknesses ranged from 10 to 20 μm. Prior to measurements, the $Sb_x Se_{1-x}$ films prepared were aged in the dark under normal laboratory conditions (i.e. $T \approx 293$ K and relative humidity 70–80%) for 2–3 weeks to allow their physical properties to equilibrate.

Antimony alloying effects on thermal generation and both deep and shallow trapping of carriers have been examined using xerographic and time-of-flight (TOF) techniques. In the xerographic measurement, the sample was charged to a potential

V_0 by passing it under a corotron (corona charging) device, which deposits charge of appropriate sign on the surface of the film. The surface potential was then measured by a transparent probe and an electrostatic voltmeter. The photoreceptor, the corona charging units and a transparent probe were housed in a well-shielded and dark environment. Following the initial charging process, the sample was exposed to strongly absorbed 450 nm step illumination from a tungsten light source, during which time the decay of the surface electrostatic potential was monitored by the electrostatic voltmeter. The surface potential decays to a potential V_r (termed the residual potential) and the resulting photoinduced discharge curve (PIDC) can be used to determine the xerographic photosensitivity of the sample. The above xerographic step could be repeated any number of times to obtain a cycled-up residual surface potential V_{rn} as a function of the xerographic cycle n.

The xerographic measurements have been complemented by TOF transient photoconductivity measurements to study the charge transport properties of $Sb_x Se_{1-x}$ alloys. The experimental TOF technique has been described in the literature by a number of authors [3, 12, 13]. In a-Se-based materials, it is more useful to measure time-resolved transits in the current mode of operation because of the dispersion and relatively long transit at low fields. In the present case we used a nitrogen gas laser to photoexcite the carriers. The nitrogen gas laser provided a short light pulse of duration $\sim$10 ns at a wavelength of 337 nm. A semitransparent gold electrode was sputtered onto the surfaces of the films as the top electrode. Because the photocarriers are generated at the sample surface owing to the high absorption coefficient of $Sb_x Se_{1-x}$ glasses at 337 nm $\left(>10^4\,cm^{-1}\right)$, the species of drifting carriers can be chosen by changing the polarity of applied voltage across the sample. The transient current was amplified and displayed on a single-event storage oscilloscope. The TOF measurement was carried out in the single-shot mode of operation, and between each measurement the sample was short-circuited and stored (relaxed) in the dark to allow any bulk space-charge build-up to decay. Small-signal conditions were maintained throughout all the measurements.

10.1 Basic Properties

First, we begin with a brief review of some fundamental physical properties. We defined the glass transition temperature T_g in the usual way, as the midpoint of the transition endotherm. Starting with 1 at% Sb, our data for T_g show a monotonic increase with increasing Sb concentration. At the same time, we observe a change of slope (step-like behaviour) between $x = 0$ and $x = 0.02$, signalling a change in the structure. For each composition we carried out several experiments, and all showed good reproducibility. Note that the experimental points represent average values, with the size of the plotted bars comparable with the size of the plotted experimental point. It should be noticed further that microhardness and density show similar compositional trends with a local rise around the same composition.

Table 10.1 The dc conductivity (at 300 K), the corresponding activation energy and the Tauc gap of $Sb_x Se_{1-x}$ glasses

Sb content [at%]	σ_{RT} $\left[\Omega^{-1}\,cm^{-1}\right]$	E_σ [eV]	dV/dt $\left[V\,s^{-1}\right]$	$t_{V_0}/2$ [s]	E_0 [eV]
0	$*1 \times 10^{-14}$	1.05	7.8×10^{-2}	267	2.04
2	$*1 \times 10^{-13}$	1.02	2.0×10^{-1}	81	2.02
5	8×10^{-11}	0.90	1.83	76	
8	9×10^{-10}	0.84			
10	4×10^{-10}	0.76			1.85
20	1×10^{-9}	0.72			1.62
40	2×10^{-8}	0.62			1.50

for any 'P'

For all Sb–Se compositions used in this study, the dark dc conductivity can be expressed in the temperature range considered by an Arrhenius-type relation

$$\sigma = \sigma_0 \exp\left(-E_\sigma/kT\right), \tag{10.1}$$

where E_σ is the activation energy and σ_0 the conductivity pre-exponential factor. The activation energy and the conductivity at 300 K are presented in Table 10.1. It can be seen that both the conductivity and the activation energy decrease with increasing Sb content.

In the glassy Sb–Se system the optical absorption edge (the Urbach tail region) shifted to longer wavelengths and the slope becomes smaller as the percentage of Sb increased. This is a general trend over most of the composition range. Absorption above the fundamental edge follows the familiar Tauc law

$$ahv = C\left(hv - E_T\right)^2, \tag{10.2}$$

where hv is the photon energy, E_T the Tauc gap and C a constant indicating how steeply the absorption rises with energy. For most compositions, there was no deviation from the square law at the highest absorption value measured in this study, exceptions being Se and films with less than 1 at% Sb. The latter follow a linear law

$$ahv = C_1\left(hv - E_1\right) \tag{10.3}$$

in the range 2.1–2.5 eV. Here C_1 is a constant and E_1 an extrapolated optical gap. The addition of less than 1 at% Sb to pure Se is enough to cause a complete transition from the anomalous linear behaviour to Tauc's law. The room-temperature values of E_T (averaged over several samples) for various $Sb_x Se_{1-x}$ compositions are given in Table 10.1. It should be noted that the Tauc gap E_T is close to that of pure Se for small Sb concentrations (<1 at% Sb) and then it decreases almost linearly up to nearly 40 at% Sb.

10.2 Dark Discharge

Typical dark discharge characteristics for pure Se and $Sb_x Se_{1-x}$ photoreceptors are shown in Fig. 10.1 for several compositions. It is apparent that for pure a-Se the decay of the surface potential is relatively slow. Comparison of the respective characteristics for a-$Sb_x Se_{1-x}$ with the dark discharge behaviour of pure a-Se shows clearly that alloying a-Se with Sb increases the dark decay rate. The discharge rates in a-$Sb_x Se_{1-x}$ were not constant but decreased with time.

There are several physical processes that can lead to the decay of the surface potential. The presently accepted model for the dark decay in a-Se-based films is that which involves [3, 14, 15]:

- Surface generation and injection of trapped electrons and their consequent transport across the sample
- Substrate injection
- Bulk thermal generation of carriers of one sign and depletion

With relatively thick films $(10 \,\mu m \le L \le 50 \,\mu m)$ and a good blocking contact between a-Se-based films and the pre-oxidised Al substrate the third phenomenon dominates.

In a series of experiments carried out on a composition series of glassy $Sb_x Se_{1-x}$ alloys, it was found that the time-dependent dark decay rate of the potential to which a dark relaxed film has been charged is controlled by the depletion discharge process. In general, the xerographic depletion discharge model is based on bulk thermal generation involving the ionisation of a deep mobility gap centre to produce a mobile charge carrier of the same sign as the surface charge and an oppositely charged ionic centre [14, 15]. Assuming negative charging, a mobile electron would be thermally generated and the ionised centre would be positive. As thermally generated holes are swept out by the electric field, a positive bulk space charge builds up with time in the specimen, causing the surface potential to decay with time.

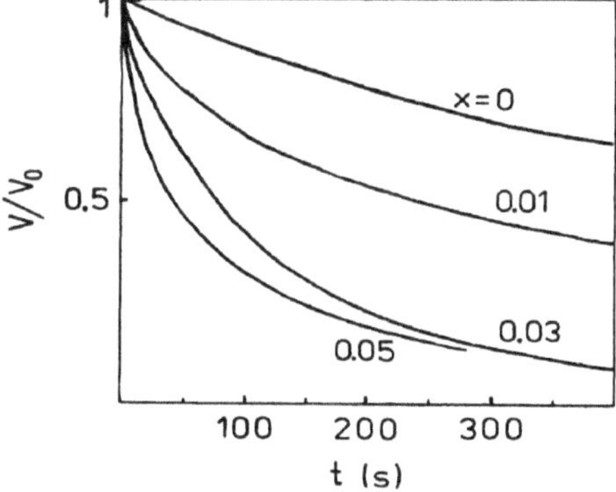

Fig. 10.1 Dark discharge of surface potential on a-SbxSe1-x layers. The surface potential at time t is normalized to that at t = 0 (initial charging potential $V_0 = 260V$).

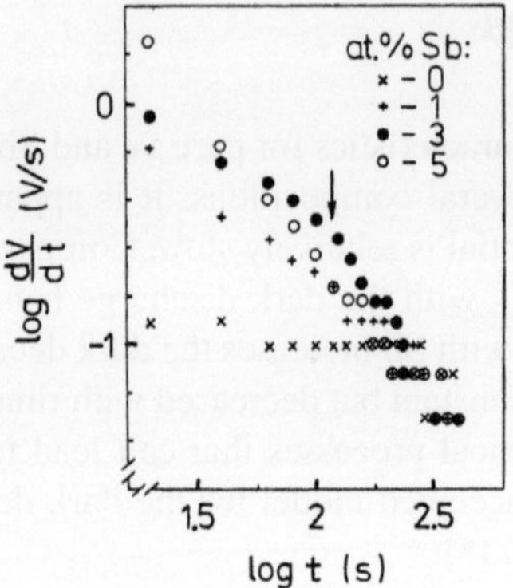

Fig. 10.2 Typical discharge rate data for SbxSe1-x films of indicated Sb content.

Figure 10.2 depicts typical dark discharge data for four different compositions, which illustrate the predicted characteristics of depletion discharge behaviour. Inflections in the log–log plots at the respective depletion times (marked by an arrow for the composition $Sb_{0.03}Se_{0.97}$) are readily identifiable. From the temperature dependence of depletion time it is estimated that the emitting sites are located $\sim 0.9 \pm 0.05$ eV below the conduction band mobility edge. Some special problems can, however, complicate the observation of a depletion kink in pure a-Se. The dark discharge rate was typically so slow in a-Se that results were always perturbed by injection. The main reasons for pure Se possessing good dark decay characteristics are (a) the remarkably small integrated number of deep localised states $\left(\sim 10^{13}\,\mathrm{cm}^{-3}\right)$ in the mobility gap of a-Se, and (b) the energy location of these states is deep ($E_t \sim 1.0$ eV) in the mobility gap, so that the thermal generation process of carriers from these centres is slow [3, 6, 15].

It is found that in a-$Sb_x Se_{1-x}$ alloys, electrons (the mobile carrier species) are depleted (n-type system) during dark decay, leaving behind a deeply trapped positive space charge. Note that the same situation prevails in alkali-doped a-Se [16].

10.3   Transient Photoconductivity

We consider first the carrier drift in pure amorphous selenium. Both the electron and hole drift mobility can be measured in a-Se by the TOF technique outlined above. At temperatures above 200 K, a well-defined transit pulse is observed. The transient profile for a well-relaxed (dark-adapted) sample is a quasi-rectangular pulse. The photocurrent remains approximately constant up to 0.2 μs and 6 μs (for holes and electrons, respectively), then decreases abruptly. The signals indicate essentially Gaussian transport. The transit time t_T was defined to correspond to the break point in the photocurrent. This represents the transit time of the fastest carriers. The mobility μ is then calculated using $\mu = L^2 / (t_T V_0)$, where V_0 is the applied voltage and L is the sample thickness. These room temperature values of the drift mobility, $\mu^h \approx 2 \times 10^{-1}\,\mathrm{cm^2\,V^{-1}\,s^{-1}}$ and $\mu^e \approx 7 \times 10^{-3}\,\mathrm{cm^2\,V^{-1}\,s^{-1}}$, are in remarkably good agreement with reported values [3, 5–7, 12, 13].

Note that for a-$Sb_x Se_{1-x}$ films deposited on a room-temperature substrate it was not possible to detect any pulses associated with the transit of hole carriers. In all samples, hole response showed a rapid decay with no apparent break. There are two plausible explanations. First, the signal is limited by the presence of deep gap states with extremely high efficiency for carrier trapping. Second, it might be argued that alloying may have increased the conductivity of the samples so that TOF experiments are no longer applicable. However, results showed that although the conductivity increases with Sb addition, it remains sufficiently low $\left(\leq 10^{-11}\,\Omega^{-1}\,\mathrm{cm^{-1}}\right)$ for TOF experiments to be applicable. It is interesting that hole response in As-alloyed a-Se has also been found to be undetectable in the range 2–4 at% As [5, 17].

The effect of Sb on electron transport is not so drastic. Although Sb alloying increases the transit time dispersion, the transit time shown contains a clearly identifiably break in the wave form. The electron drift mobility in a-$Sb_x Se_{1-x}$ alloys exhibits Arrhenius behaviour. The experimentally observed activation energy of a-Se, namely $E_\mu \approx 0.33 \pm 0.01\,\mathrm{eV}$, remains almost insensitive to Sb addition. At the same time, the mobility decreases with increasing Sb content. It is reasonable, within the framework of a shallow-trap-controlled mobility model [18], to interpret our TOF observations in the following manner. In pure Se, electron transport is controlled by a narrow manifold of traps located about $0.33\,\mathrm{eV}$ from the conduction band mobility edge. The addition of Sb to Se broadens the distribution of shallow traps, we believe, thus increasing the relative dispersion of photocurrent transients. The decrease in electron mobility with increasing Sb content can be accounted for by the increase in the density of shallow traps with E_μ remaining constant at approximately $0.33\,\mathrm{eV}$.

10.4 Photoinduced Discharge Characteristics

In essence, the xerographic photosensitivity S of a photoreceptor material determines the rate of decay dV/dt of the electrostatic surface potential of the samples during photoinduced discharge (PID). The xerographic photosensitivity definition adopted here is simply based on the amount of light energy required for the surface potential to decay to half of its original value $(V_0/2)$ during photoinduced discharge (i.e. the fractional change in the surface potential per unit light exposure).

Figure 10.3 shows a dark decay curve and a photoinduced discharge curve for the a-$Sb_{0.03}Se_{0.97}$ film. It can be seen that the sample exhibits relatively little dark decay. Nevertheless, in order to take into account the surface charge reduction during illumination and to evaluate S accurately, the contribution of dark discharge to the total change in the surface potential during PID was subtracted ($V_0 - V_d$ in Fig. 10.3).

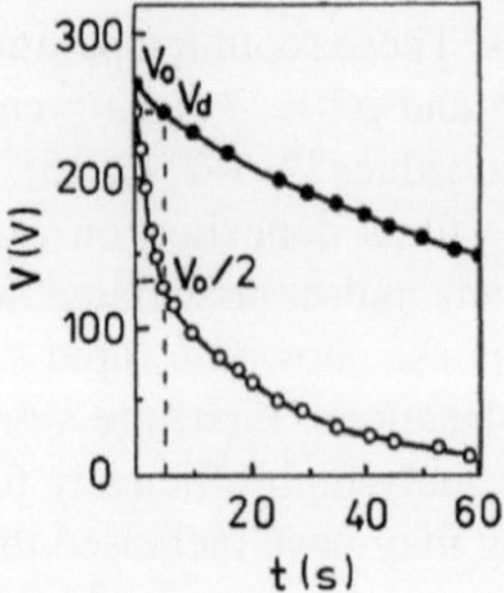

Fig. 10.3 Dark decay curve (solid circles) and photoinduced discharge curve (open circles) for $Sb_{0.03}Se_{0.97}$.

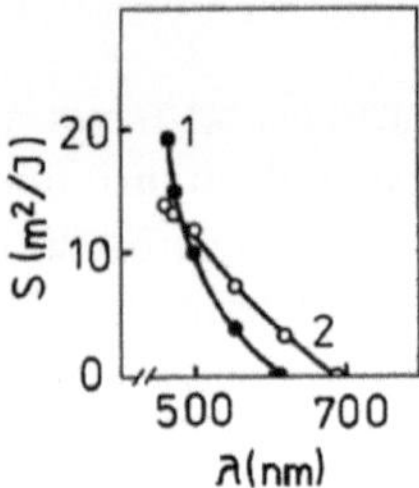

Fig. 10.4 Xerographic photosensitivity (S) versus exposure wavelength for pure Se (1) and $Sb_{0.03}Se_{0.97}$ alloy films. For all measurements the light intensity was kept constant at 0.7 J/m s and electric field was kept equal at 23000 V/cm.

The xerographic spectral response for both a-$Sb_{0.03}Se_{0.97}$ and pure Se are shown in Fig. 10.4. It is apparent that as Sb is added to a-Se, the photosensitivity at a particular wavelength increases. More precisely, this means that the xerographic photosensitivity for Sb_xSe_{1-x} alloy is somewhat greater at longer wavelengths ($\lambda > 670$ nm) than for pure Se, and smaller at shorter wavelengths ($\lambda < 500$ nm). Note that other compositions of Sb–Se alloy showed similar trends. The result of increased xerographic photosensitivity of the Sb–Se alloys at longer wavelengths suggests that the addition of Sb to a-Se causes a reduction of the bandgap of the material. Note that the xerographic photosensitivity depends not only on the absorption coefficient α but also on the quantum efficiency η for generating mobile charge carriers, as well as the transport properties (the product of drift mobility and lifetime $\mu\tau$) [3].

For the films under examination the residual voltage V_r (a measurable surface potential at the end of the illumination) increases with Sb content (Fig. 10.5). The residual potential is due to trapped electrons in the bulk of the specimen. The simplest theoretical model, which is based on range limitation and weak trapping ($V_r << V_0$), relates V_r to $\mu\tau$ by the Warter equation [19]

$$V_r = L^2 / (2\pi\tau),$$ (10.4)

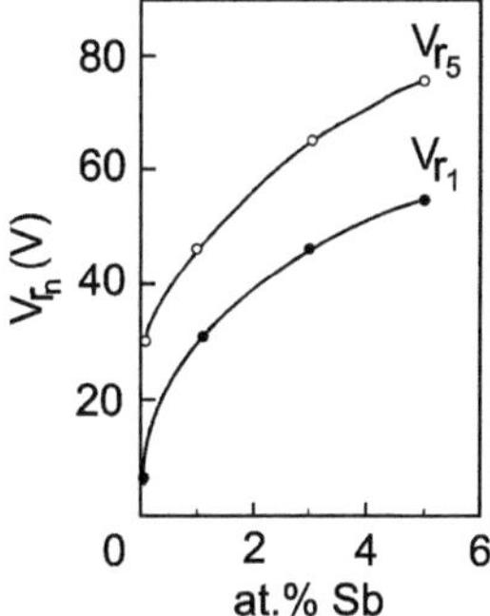

Fig. 10.5 Effect of antimony on the residual potential (measured after first and fifth cycles) of Sb-Se alloy.

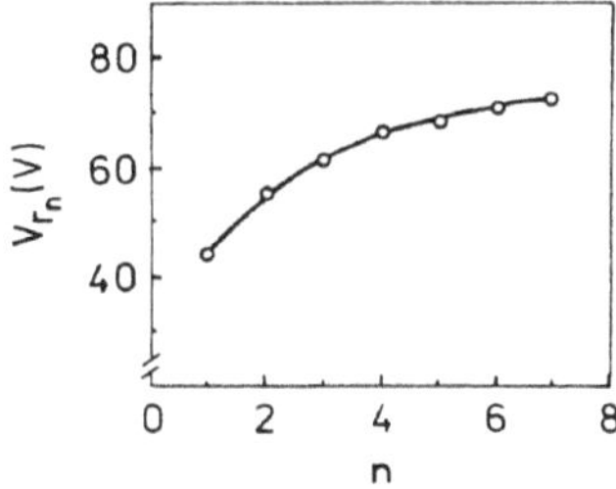

Fig. 10.6 The build-up in the residual voltage with number of xerographic cycles in a-$Sb_{0.03}Se_{0.97}$ films. One charge-discharge cycle consists of 60-s light decay with 1000 lx tungsten lamp after corona charging.

where L is the sample thickness. For example, addition of 3 at% Sb leads to a change in the first cycle residual voltage from 4 to 44 V, which is equivalent to a change of the carrier range $\mu\tau$ from 10^{-7} to 10^{-6} cm^2 V^{-1}. Substituting $\mu^e \approx 7 \times 10^{-3}$ cm^2 V^{-1} s^{-1} for pure Se and $\mu^e \approx 6 \times 10^{-4}$ cm^2 V^{-1} s^{-1} for $Sb_{0.03}Se_{0.97}$ into the corresponding equation, we find carrier lifetimes $\tau \approx 2 \times 10^{-4}$ s and $\tau \approx 1.3 \times 10^{-3}$ s in a-Se and a-$Sb_{0.03}Se_{0.97}$. Note that, in general, bulk deep trapping lifetimes computed from the first-cycle residuals agree with lifetimes measured in the TOF mode under range-limited conditions.

Figure 10.6 displays the build-up of the residual voltage V_{rn} on a-$Sb_{0.03}Se_{0.97}$ film with the number n of xerographic cycles. The rate of increase of V_{rn} decreases with number of cycles. Then, for large n (≥ 6 in our case), V_{rn} tends to a saturation value V_{rs}. As described earlier [20], the saturation residual potential provides an experimental measure of the integrated number of deep traps (trap-release rates are much slower than those from shallow traps, which control drift mobility). Then V_{rs} is simply given by

$$V_{rs} = eN_t L^2 / (2\varepsilon), \tag{10.5}$$

where N_t is the deep-trap concentration and ε the dielectric constant.

Both the first residual and the cycled-up saturated residual potential are sensitive to alloying. For example, when pure amorphous Se films are alloyed with

Sb, the build-up of the residual potential occurs more rapidly towards a much higher saturated residual potential. We obtain, for instance, $N_t \sim 2 \times 10^{14}\,\mathrm{cm}^{-3}$ and $N_t \sim 10^{15}\,\mathrm{cm}^{-3}$ for a-Se and $\mathrm{Sb}_{0.03}\mathrm{Se}_{0.97}$, respectively.

The above photoelectric properties of a-$\mathrm{Sb}_x\mathrm{Se}_{1-x}$ alloys can be at least qualitatively explained by using concepts based on charged structural defects, called valence alternation pairs (VAPs) or intimate valence alternation pairs (IVAPs). These correspond to some of the chalcogen atoms being under- and over-coordinated [21]. It seems reasonable that dark discharge and residual voltage build-up involve essentially the same species of localised centres. Further, it is also possible that we may be observing amphoteric behaviour by IVAPs. For pure Se, an IVAP comprises over- and under-coordinated selenium atoms Se_3^+ and Se_1^- in close proximity. An IVAP centre would be seen as a "neutral trap" by the carrier. The capture of an electron by Se_3^+ exposes the negative charge on Se_1^- and explains the residual voltage detected. At the same time, the emission of an electron from the Se_1^- uncovers the positive charge on Se_3^+, which causes the dark decay. Although the question remains whether the neutral centre is a D^0-type defect or whether it is an IVAP defect, the measured radius of 3 Å in [22] is representative of an IVAP capture radius.

There are a number of desirable electrical characteristics that a useful photoreceptor should exhibit. Our experimental results show that dark decay rate, residual voltage and drift mobility are all sensitive to Sb. Carrier drift mobility decreases with Sb whereas dark decay rate and residual voltage increase. All these effects are undesirable in xerography. At the same time, the main advantage of a-$\mathrm{Sb}_x\mathrm{Se}_{1-x}$ is that its spectral response can be readily shifted to longer wavelengths by increasing the Sb content, which allows for photoreceptor designs that can respond to a variety of illumination spectra. In addition, it may be possible to improve the charge transport parameters by halogen doping. Indeed, Cl is known to compensate for As-induced (As is in the same group as Sb) deep traps in a-Se [23]. By using a double-layer photoreceptor consisting of a thin a-$\mathrm{Sb}_x\mathrm{Se}_{1-x}$ layer for photogeneration and a thick a-Se layer for charge transport, it is possible to improve the xerographic parameters further.

Thus, the mobile carrier species controlling the xerographic depletion discharge in a-Sb–Se alloys are electrons. Thermal generation of free electrons in a-$\mathrm{Sb}_x\mathrm{Se}_{1-x}$ is accompanied by the simultaneous formation of deeply trapped positive space charge. It has been shown that Sb alloying progressively enhances the free electron thermal generation rate relative to the pure specimen.

As apparent from the large xerographic residual potentials for $\mathrm{Sb}_x\mathrm{Se}_{1-x}$ alloys, the addition of Sb to a-Se seems to greatly increase the concentration of deep localised states within the mobility gap of the material.

The results indicate that in the long-wavelength region (e.g. $\lambda \sim 600\,\mathrm{nm}$) the photosensitivity of the a-$\mathrm{Sb}_x\mathrm{Se}_{1-x}$ films is higher than of pure selenium, probably owing to higher quantum efficiency.

In such an exotic field of materials science as amorphous (disordered) solids, one of the fundamental problems that has been studied extensively for a long time is how to obtain insight into the structure. At present, it seems that versatile studies are needed to elucidate amorphous structure. In other words, in addition to various direct and indirect structural techniques (X-ray diffraction, Raman scattering,

IR-absorption, EXAFS, to name but a few) performed under fixed conditions, investigations of structural modifications introduced by changes in composition, temperature or pressure or induced by bandgap illumination may provide fruitful ideas.

One of the properties of the class of materials known as chalcogenide glasses is that they exhibit a wide spectrum of photoinduced effects. Photoinduced phenomena have been extensively studied recently (see the corresponding references in previous chapters), partly as an interesting subject for fundamental research in the field of disordered solids and partly due to potential application of these phenomena in opto(photo)electronics (xerography and xeroradiography, optical memories, optical circuits, photoresists, etc.).

Among these phenomena, so-called reversible photodarkening (PD) and photocrystallisation are the most interesting. The changes in various physical and chemical properties of chalcogenide glasses under bandgap illumination have been detected. It has been known since 1968 that bandgap illumination of amorphous selenium films increases the growth rate of crystallites [24]. This phenomenon was later utilised by others to develop images on 150–200 μm thick selenium layers on gold. Although several mechanisms of PD and photocrystallisation have been proposed, the details of these apparently simple phenomena remain ambiguous.

In the present section, we will also examine some features of the above two phenomena – room-temperature reversible PD and photocrystallisation – in amorphous semiconducting films of $Sb_x Se_{1-x}$. Pure amorphous Se was chosen as the starting material. As an elemental amorphous material, Se may be extremely suitable for discussing essential features and the relationship (if any exists) between the above phenomena. In addition, the effect of small amounts of Sb (a few per cent) on PD and photocrystallisation of a-Se is especially interesting not only from the point of view of compositional disordering, but also because of desirable recording properties and particular features of electronic transport for amorphous $Sb_x Se_{1-x}$ films [25].

The samples investigated were 0.3–3.0 μm amorphous $Sb_x Se_{1-x}$ $(0 \leq x \leq 0.05)$ films. These were preferentially prepared by conventional vacuum evaporation onto room-temperature silica–glass substrates (plates). The a-$Sb_x Se_{1-x}$ source material was made by the usual melt-quenching technique. The cooling rate was estimated to be 100–200 $K\,s^{-1}$. Prior to measurements, the films prepared were aged at laboratory conditions (natural aging) for several weeks to allow their structure to equilibrate.

Three kinds of measurements were performed.

- Transmission PD experiments, in which the samples were illuminated at near normal incidence by a He–Ne laser operating at 633 nm. The transmission of the samples was probed using a portion of relatively low ($\sim$3 mW) intensity of the He–Ne laser output and detected by a photomultiplier. In this experiment, the inducing and probing light propagated in parallel. The transmission was measured as a function of exposure time and intensity as well as sample composition.
- Holographic experiments. The grating technique used is a conventional method: a grating is produced by two interfering beams intersecting at a sample surface.

The measured parameter η, the diffraction efficiency, is the ratio of the corresponding intensities of the probe beam I_0 and the first-order diffracted one I.

- Structural probes. Right-angle Raman spectra and X-ray diffraction were measured at room temperature.

Photodarkening and photocrystallisation of a-$Sb_x Se_{1-x}$ films was induced by linearly polarised light with a wavelength of 633 nm emitted from He–Ne laser. The light intensity was varied from $0.5\,W\,cm^{-2}$ to $2.5 \times 10^2\,W\,cm^{-2}$.

10.5 Optical Properties

Figure 10.7 shows the change in transmissivity as a function of exposure time for amorphous $Sb_x Se_{1-x}$ films. The relative transmissivity T_{rel} is defined as T_{ir}/T_{un}, where T_{ir} and T_{un} are for the irradiated and unirradiated samples, respectively. One can clearly see a decrease in T_{rel} with the irradiation time. Light-induced change of transmissivity is transient: the initial value of T_{rel} is restored by switching off the light. This is not surprisingly because the glass-transition temperature of the compositions examined is located at approximately room temperature (pure a-Se) or slightly above this value (Se containing < 5 at% of Sb additives). Note that light with low intensity $\left(I < 10\,mW\,cm^{-2}\right)$ could not induce appreciable permanent effects, despite long exposure times. Only transitory behaviour in T_{rel} is observed.

On the other hand, with increasing exposure (more precisely, its intensity) significant irreversible changes in transmissivity are observed (Fig. 10.8). We see that the characteristics below and above the threshold intensity I_{th} are quite different. In general, the T_{rel} versus E curve, where E is the exposure magnitude, may be

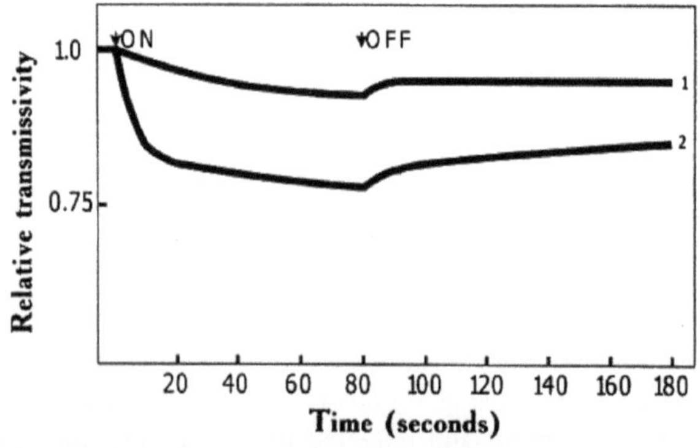

Fig. 10.7 Transitory changes in transmissivity in a-$Sb_x Se_{1-x} Sb_x Se_{1-x}$ films caused by switching on and off irradiation of 633 nm wavelength. Film thickness 0.8 μm, illumination intensity $0.5\,W\,cm^{-2}$, Sb concentration 1 (*curve 1*) and 5 at% (*curve 2*). On- and off-periods of illumination are shown by *arrows*

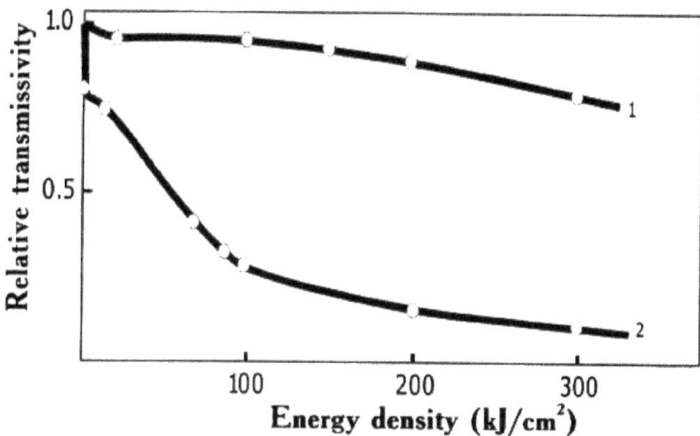

Fig. 10.8 Relative transmissivity versus energy density in amorphous $Sb_xSe_{1-x}Sb_xSe_{1-x}$ films exposed to intense laser illumination at 633 nm. Film thickness 0.8 μm, intensity 2.5×10^2 W cm^{-2}, Sb concentration for curves 1 and 2 is 1 and 5 at%, respectively

divided into three parts:

- Initial, with a rapidly varying response, a transient transmission change
- Middle, slowly varying portion of the darkening curve; sometimes this part is marked by the beginning of a plateau region for intensities $I \approx I_{th}$
- Final, with a significant (up to 0.8–0.9) decrease in T_{rel}, a permanent, irreversible change

The photoeffect is completed by a saturation region; as above, the onset of the latter is intensity dependent.

Changes in transmissivity for $I \geq I_{th}$ may be attributed to crystallisation transformation (see below).

Nearly the same behaviour can be discerned for holographic recording properties. The η versus E data also exhibit several stages depending on energy density. Typical responses of light intensities diffracted from gratings formed on films are shown in Fig. 10.9 (low intensities) and Fig. 10.10 (high intensities): an initial increase to $\eta_{max} = 0.012\%$, a slight decrease then a monotonic increase with E leading to the maximum $\eta = 6\%$. The high energy density side of the above maximum is caused probably by optical absorption of the probing beam increasing due to photocrystallisation.

The degree of change in T_{rel} and η increases with Sb content (see Figs. 10.7, 10.8, 10.10). Apart from this, the addition of Sb to a-Se shifts the crystallisation onset to higher exposure values.

10.6 Structural Transformation

Figure 10.11 shows the evolution of Raman spectra with increasing light intensity of illumination. It is important to point out here the existence of a certain threshold intensity of the incident laser beam: below I th the initial shape of the spectrum was

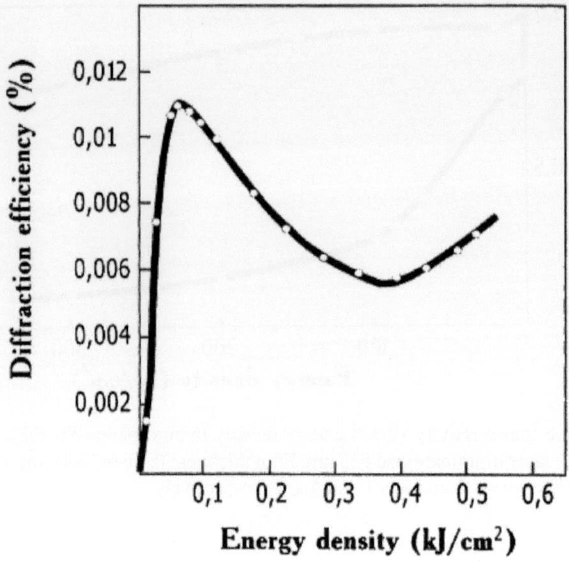

Fig. 10.9 Diffraction efficiency versus energy density for pure a-Se in the low energy density region. The *line* is to guide the eye

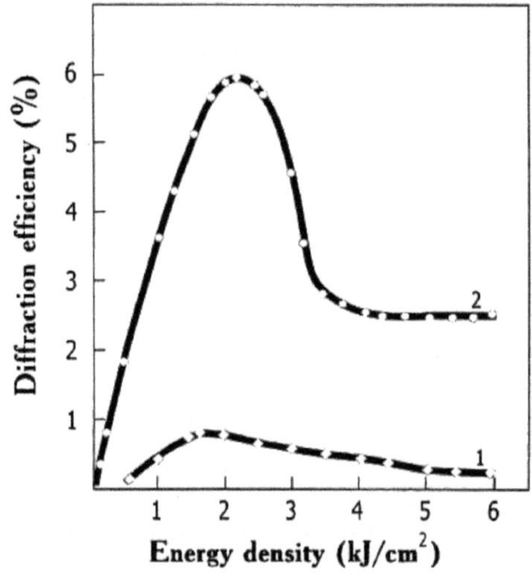

Fig. 10.10 Changes of diffraction efficiency in a-Sb$_x$Se$_{1-x}$Sb$_x$Se$_{1-x}$ samples induced by laser irradiation. Curves 1 and 2 correspond to $x = 0$ and 0.03, respectively. Film thickness $1.25\,\mu$m, intensity $1.25\,$W cm^{-2}

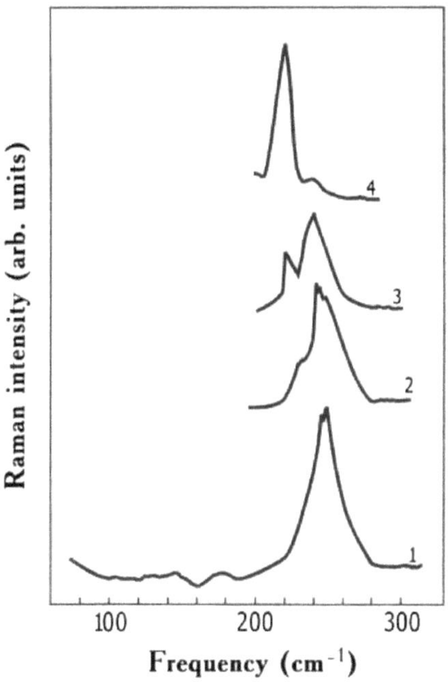

Fig. 10.11 Transformation of the Raman spectrum in amorphous $Sb_{0.03}Se_{0.97}$ exposed to linearly polarised light: curve 1, reference Raman spectrum of amorphous state; 2–4, after exposure to $E = 3, 5$ and $6\,kJ\,cm^{-2}$, respectively. The inducing intensity was $I = 1.25\,W\,cm^{-2}$

recovered after turning off the illumination, above it the spectrum further transforms (spontaneously) even in the absence of excitation at this stage. Note that the samples illuminated at $E < E_{th}$ are amorphous, while the samples illuminated at $E > E_{th}$ exhibit well-pronounced crystalline features in the Raman spectra.

Some comments on the Raman spectra are called for here. First, the spectra of the amorphous Sb_xSe_{1-x} samples before irradiation are close to that for pure Se. The only additional feature observed is the appearance of the $\sim 190\,cm^{-1}$ band of $Sb_2Se_{3/2}$ structural units. Second, the exposure values at which the spectra start to transform vary with addition of Sb. Third, under intense irradiation, Raman spectra show a gradual intensity redistribution for the peaks at $237\,cm^{-1}$ and $255\,cm^{-1}$. When the radiation power is further increased, the intensity of the $237\,cm^{-1}$ peak grows continuously with a simultaneous decrease of the $255\,cm^{-1}$ spectral feature. Finally, it appears that Sb addition in the quantity used here (<5 at%) has no appreciable influence on photocrystallisation product. Actually, only the $237\,cm^{-1}$ Raman band of hexagonal Se contributes to the spectra of photocrystallised Sb_xSe_{1-x} films. Thus, it is indicated clearly that on introducing a small quantity of additives to Se glass, there is no appreciable influence on the crystallisation product. This can be explained in an unambiguous way: the small quantity of Sb is clustered in the glass matrix. Further, in the case when the sample was illuminated, Se crystallised out, while the additives still remained in the disordered (glassy) state. Even though a higher or lower E_{th} is found in samples of different chemical composition, there is little influence on the kind of crystallised product. In addition, it should be noted that evolution of the shape of Raman spectra similar (at least qualitatively) to that presented here has also been observed for amorphous $As_x Se_{1-x}$ alloys over the concentration range 0–10 at% As.

X-ray measurements also indicate photocrystallisation at exposure values $E > E_{th}$. Typical results, that is, X-ray diffraction patterns of photocrystallised samples (i.e. samples corresponding to curve 5 of Fig. 10.11), show four crystalline peaks, located at $2\theta = 24°$, $30°$, $41°$ and $45°$. These can be indexed as 100, 101, 110 and 111 peaks for the hexagonal Se crystal. Similar X-ray diffraction patterns were obtained by Ishida and Tanaka [26] on examining photoinduced birefringence in a-Se thin films.

The present experimental results permit three successive stages of laser-induced changes in a-$Sb_x Se_{1-x}$ to be distinguished with regard to the irradiation energy density.

Stage I. First, the initial stage: dynamical (transient) photoinduced effects in T_{rel}, η, etc. Selenium, as a chalcogen, exhibits reversible PD. However, the phenomenon has not been studied in detail. The reason for this is that permanent (quasi-stable) photoinduced changes can only be realised at low $(T < 200\,\mathrm{K})$ temperatures – a complete recovery of initial parameters was observed for samples annealed at room temperature. It is essential to note that the sample remains in the *amorphous* state before, during and even after irradiation. Reasonably, one may argue that the observed dynamical (transitory) change in transmissivity under light irradiation may be caused by thermal excitation (in other words, it is thermal in origin). Nevertheless, because of the following arguments, we believe this is not a dominant factor, although some heating up of the sample cannot be fully eliminated.

- The above photoeffect shows no dependence on the substrate material (note that we examined several types of substrate: poly(methylmethacrylate), quartz glass, mica foils, etc.). It is evident that for the substrate materials mentioned the heat dissipation conditions are significantly different.
- The lack of any noticeable variation in T_{rel} (or η) behaviour for samples of different film thickness significantly reduces the possibility of the effect being due to changes in temperature during illumination.

Some photoelectronic properties exhibit similar, dynamic, photoinduced changes [27]. We relate the transient changes in the transmissivity (PD) to changes in metastable deep states in the mobility gap. The states are associated with charged structural defects (e.g. IVAPs), which correspond to some of the chalcogen atoms being over- and under-coordinated. For pure Se, an IVAP comprises Se_3^+ and Se_1^- in close proximity. Band-gap illumination initiates conversion of such states into ones with greater density or capture cross section.

Stages II and III. The intermediate and final stages are both associated with crystallisation: the former with microcrystallite formation, the latter with microcrystallite enlargement and their concentration (nucleation and growth of crystalline structures).

In order to discuss an essential feature of the laser-induced structural change, we first consider the structure of amorphous films. Almost all available structural data indicate that the main constituent of a-$Sb_x Se_{1-x}$ ($x < 0.05$) is the chain molecule, though molecules with Sb branching sites may be present in a minority. That is, we assume that the local structure of a-Se containing Sb additives resembles the hexagonal (trigonal) Se structure. Accordingly, in atomic structural terms, a-$Sb_x Se_{1-x}$ with $x < 5$ at% may be characterised as a quasi-one-dimensional chain structure.

Plausible explanations for the present experimental observations can provide the quasicrystalline model proposed by Zhdanov et al. [28], Shimakawa et al. [29], Tanaka and Ishida [30] and Fritzsche [31]. Following Fritzsche's phenomenological idea, we assume that the amorphous structure of annealed and then dark-rested films contains some anisotropic elements. These are aligned in random orientation, so the structure appears to be isotropic, a characteristic inherent to amorphous solid-state materials. If the above amorphous samples are exposed to linearly polarised light, structural elements (chain segments) tend to align in some preferential orientation, which is nearly perpendicular to the electric field vector of linearly polarised light. Because Se-rich alloys in their non-crystalline form can be regarded as a kind of a "soft" material (its atomic structure is flexible due to twofold coordination of chalcogen atoms), chain-segment orientation can be influenced relatively easily. The intensity difference between the horizontal and vertical configurations of the 101 peak observed in a series of X-ray diffraction patterns seems to be consistent with the preferential alignment of the chain molecules perpendicular to the electric field vector of linearly polarised light. Accordingly, in spite of this microcrystalline model, if the crystallisation process proceeds in the manner described above, and if the structural elements mentioned are responsible for the photoinduced anisotropy in the amorphous state, we may speculate that such elements with quasicrystalline orientation induced in amorphous phase by linearly polarised illumination of low intensity (referred to as the *initial* stage) become nuclei, which grow as oriented crystals. The energy density needed for the latter process should be greater than the threshold energy E_{th}. Although microscopic structures giving rise to the light-induced transitory changes and photocrystallisation still remain controversial, it is assumed that the latter phenomena arise from changes in structures at scales more extended than the atomic scale.

Finally, it is interesting to consider the above results in relation to other Se-based binary alloys containing group V elements as a constituent. For example, the same behaviour is observed for the $As_x Se_{1-x}$ system when the latter is subjected to band-gap illumination. Moreover, we find that many of the physical properties (glass transition temperature T_g, density d, bandgap energy E_0, etc.) of the $Sb_x Se_{1-x}$ system change around $0.01 \leq x \leq 0.02$. The same behaviour is observed for $As_x Se_{1-x}$, which has local turning points (inflections) in several properties below the usual threshold, which is near the composition $x \approx 0.04$. A possible explanation suggested is that the peculiarities mentioned are associated with the topological threshold that occurs when the chain-ring-like structure changes to a chain-like structure [32, 33].

Very important is the observed similarity between the light-induced changes in pure amorphous Se and Se containing Sb or As additives. Probably, the similarity is the clue to a more general understanding of various photoinduced changes, photocrystallisation and compositional dependence in chalcogenide vitreous semiconductors. Further experiments would be very helpful in this context.

Based on the above results we can say that linearly polarised light from the bandgap absorption region can induce either transitory changes or crystallisation transformation in amorphous Sb_xSe_{1-x} alloys. These two phenomena critically depend on exposure, show threshold behaviour and seem to arise from mechanisms apparently different: defect states or some kind of structural units liable to preferential orientation under the action of linearly polarised light.

References

1. A. Madan, M.P. Shaw, *The Physics and Applications of Amorphous Semiconductors* (Academic, Boston, MA 1988)
2. K. Tanaka, Curr. Opin. Solid State Mater. Sci. **1**, 567 (1996)
3. S.O. Kasap, in *Handbook of Imaging Materials*, ed. by A.S. Diamond (Marcel Dekker, New York 1991) p. 329
4. J. Rowlands, S. Kasap, Phys. Today **50**(11), 24 (1997)
5. J. Schottmiller, M. Tabak, G. Lucovsky, A.I. Ward, J. Non-Cryst. Solids **4**, 80 (1970)
6. M. Abkowitz, J.M. Markovich, Solid State Commun. **44**, 1431 (1982)
7. N.F. Mott, A.E. Davis, *Electronic Processes in Non-Crystalline Materials* (Clarendon, Oxford 1979)
8. C. Wood, R. Mueller, L.R. Gilbert, J. Non-Cryst. Solids **12**, 295 (1973)
9. M.H. El-Zaidia, A. El-Shafi, A.A. Ammar, M. Abo-Ghozala, Thermochim. Acta **116**, 35 (1987)
10. F.V. Pirogov, J. Non-Cryst. Solids **114**, 76 (1989)
11. V.I. Mikla, Phys. Status Solidi B **182**, 325 (1994)
12. W.E. Spear, J. Non-Cryst. Solids **1**, 197 (1969)
13. G. Pfister, H. Scher, Adv. Phys. **27**, 747 (1978)
14. A.R. Melnyk, J. Non-Cryst. Solids **35/36**, 837 (1980)
15. M. Abkowitz, F. Jansen, A.R. Melnyk, Philos. Mag. B **51**, 405 (1985)
16. M. Abkowitz, F. Jansen, J. Non-Cryst. Solids **59/60**, 953 (1983)
17. V.I. Mikla, Mater. Sci. Eng. B **49**, 123 (1997)
18. J.M. Marshall, F.D. Fisher, A.E. Owen, Phys. Status Solidi A **25**, 419 (1974)
19. P.J. Warter, Appl. Opt. Suppl. **3**, 65 (1985)
20. M. Abkowitz, R.C. Enck, Phys. Rev. B **27**, 7402 (1983)
21. M. Kastner, D. Adler, H. Fritzsche, Phys. Rev. Lett. **37**, 1504 (1976)
22. B. Polischuk, S.O. Kasap, V. Aiyah, M. Abkowitz, Can. J. Phys. **69**, 364 (1991)
23. S.M. Vaezi-Nejad, C. Juhasz, J. Mater. Sci. **23**, 3286 (1988)
24. J. Dresner, G.B. Stringfellow, J. Phys. Chem. Solids **29**, 303 (1968)
25. F.V. Pirogov, J. Non-Cryst. Solids **114**, 76 (1989)
26. K. Ishida, K. Tanaka, Phys. Rev. B **56**, 206 (1978)
27. V.I. Mikla, Yu. Nagy, V.V. Mikla, A.V. Mateleshko, Mater. Sci. Eng. B **64**, 1 (1999)
28. V.G. Zhdanov, B.T. Kolomiets, V.M. Lyubin, V.K. Malinovskii, Phys. Status Solidi A **52**, 621 (1979)
29. K. Shimakawa, A. Kolobov, S.R. Elliott, Adv. Phys. **44**, 475 (1995)
30. K. Tanaka, K. Ishida, J. Non-Cryst. Solids **227**, 673 (1998)
31. H. Fritzsche, Phys. Rev. B **52**, 15854 (1995)
32. V.I. Mikla, J. Phys. Condens. Matter **9**, 9209 (1998)
33. S.O. Kasap, D. Tonchev, T. Wagner, J. Mater. Sci. Lett. **17**, 1809 (1998)

Index